MORE PRAISE FOR *TINKERING WITH EDEN*

"[An] intriguing history. . . . An essential look at the unexpected and, all too often, unwelcome impact exotics have on the ecosystem in which they thrive." —*Booklist,* "Top 10 Sci-Tech Books"

"Todd masterfully weaves details of animal behaviors and plant ecology gleaned from the scientific literature with details from historical records to create fascinating and readable accounts of species introductions." —*Choice*

"By weaving together natural science, political history, and intriguing profiles of the people who introduced exotic animals with the best of intentions—and often with disastrous results—Todd offers an entertaining, enlightening, and important new history." —*Utne Reader*

"Kim Todd's *Tinkering with Eden* provides a clear, objective harbinger of how much the introduction of exotic species has changed the landscape since Christopher Columbus first set foot this side of the Atlantic." —*Austin Chronicle*

"A fascinating narrative enhanced by Todd's far-reaching research and rich story-telling abilities. . . . Her book will interest any caring observer of our environment or lover of mystery." —*BookPage*

"An instructive and remarkably fresh look at non-native species. . . . Witty and lyrical essays." —*Express Books*

"The lively tales compel fascinated attention. . . . Kim Todd has a gift for explaining science." —*The News & Observer* (Raleigh, North Carolina)

"Todd uncovers a Greek tragedy of human heedlessness. . . . [A] beautifully written natural history." —*Outside*

TINKERING WITH EDEN

A NATURAL HISTORY OF

EXOTIC SPECIES IN AMERICA

KIM TODD

ILLUSTRATIONS BY
CLAIRE EMERY

W · W · NORTON & COMPANY

NEW YORK   LONDON

Copyright © 2001 by Kim Todd

For information about permission to reproduce selections from this
book, write to Permissions, W. W. Norton & Company, Inc.,
500 Fifth Avenue, New York, NY 10110

The text of this book is composed in 11/13.5 Adobe Garamond
with the display set in Sanvito Light
Composition by Sue Carlson
Manufacturing by The Haddon Craftsmen, Inc.
Book design by Margaret M. Wagner
Illustrations by Claire Emery

Library of Congress Cataloging-in-Publication Data

Todd, Kim, 1970–
    Tinkering with Eden : a natural history of exotics in America /
Kim Todd.
        p. cm.
    Includes bibliographical references and index.
    ISBN 0-393-04860-8
    1. Animal introduction. I. Title.
    QL86.T64 2001
    591.6—dc21   00-058740

ISBN 0-393-32324-2 pbk.

W. W. Norton & Company, Inc., 500 Fifth Avenue,
New York, N.Y. 10110
www.wwnorton.com

W. W. Norton & Company Ltd., Castle House, 75/76 Wells Street,
London W1T 3QT

2 3 4 5 6 7 8 9 0

FOR JAY

CONTENTS

III. HERE AND NOW

Versions of the chapter titled "Words on the Wing" appeared earlier under the title of "Starling" in *Orion* and the *Bellingham Review*. Reprinted with permission from *Orion*.

THE ultimate elegance: the imagined land

—WALLACE STEVENS

WHAT you are telling us is that there are some
bad critters which we know in South Texas as the
Fire Ant, the Boll Weevil, killer bees, probably
even nutria which was introduced for good cause
and has created interest. Then you come along
and tell us there really are some wonderful things
that are non-indigenous. How do we in this
Committee determine what is good or what is
bad? Do you have a procedure?

—REPRESENTATIVE GREG LAUGHLIN
of Texas in response to the Office of Technology
Assessment's *Report on Introduction of Harmful
Non-Indigenous Species into the United States* in 1993

TINKERING WITH EDEN

INTRODUCTION

CONJURE up a landscape. On the cusp of city and country. A place where birds and cars make equal noise. Somewhere typically American. Stroll by one of the last houses on the edge of town, the one with the messy lawn. Dandelions poke through long stalks of Kentucky bluegrass. House sparrows fluff brown feathers in the hedge, and starlings probe the soil with yellow beaks. Flower gardens give way to fields, and a calf stumbles to its feet in the tall timothy. Honeybees fumble at nodding thistles. Farther out, as the road turns to dirt, a ring-necked pheasant flushes out of the bushes, croaking, sun catching his iridescent neck. Brown trout arc and break the surface of a stream leading into the woods.

Though the scene might look familiar, glance back. None of the species that swim, fly, or take root in this dreamscape is native to the United States. They evolved in Asia, Africa, or Europe, shaped by foreign rivers and grasslands, and have managed to thrive on this far continent. Since Europeans began settling the area we now call America, exotic species have flooded in, becoming so prevalent that many Americans can't say which plants and animals are native and which not. While often invisible to the naked eye, changes in landscapes and ecosystems develop as these species take up residence and begin to breed. In many areas of San Francisco Bay, where trade ships have docked for centuries, 99 percent of all life is exotic. Exotics may have contributed to the decline of 49 percent of threatened and endangered species. On the other hand, as Alfred Crosby points out in *Ecological Imperi-*

alism, other nonnative species are inextricably tied to the way most of us live, and their homey presence permitted the success of English, Spanish, and French colonists in this land pulsing with strange beasts. Currently more than forty-five hundred exotic creatures buzz, creep, and wing their way over American soil, and each has a very specific, compelling history.

While many newspaper and magazine articles point out the disasters unleashed with these species, few look deeply at why people wanted to release them in the first place. Many of the animals came over in boats traveling from the Old World. While colonists brought germs in their lungs and rats on their ships unintentionally, they carried other creatures here by design, following an impulse as old as Noah's Ark, to re-create a remembered country across the sea. Settlers in New France shipped over rock doves to keep as pets, little imagining the hordes of pigeons that would grow with the country's cities. In 1890 Eugene Schieffelin, a Manhattan drug manufacturer, went on a crusade to release European birds into Central Park, and the starlings he set free are the ancestors of those that sit clicking and squawking on telephone wires from Miami to Spokane. Newspaper editor E. B. Webster transported mountain goats to the Olympic Peninsula, and astronomer Leopold Trouvelot brought gypsy moths to Massachusetts. These men were looking to adjust the world as they found it, to tinker with the Garden of Eden. They had visions, and they wanted to build them out of cells rather than stone. As new technology made transportation easier and quicker, more and more visionaries picked animals out of one environment and set them free in another. Why?

It is a simple question with a complicated answer. Transporting enough animals to launch a population generally required either wealth or a government job, particularly early on, so reasons spanned from the public good to private whim. People brought species here to provide food, keep them company, remind them of home, attract tourists, and spur good hunting.

Some men were keeping pace with their times, reflecting cultural desires. Others regarded exotic species as winning lottery tickets, as the means to make their fortunes through silk or furs. Still others were renegades, fueling their own dreams as they unlatched the cages.

Ultimately these releases say less about the animals themselves than about how we imagine our relationship with other species. In its own fashion each introduction underscores the dilemma of humans and the natural world: In some ways we have complete control; in others none. The urge to release a few pairs of chukar partridge can trigger a biological explosion, but all our science cannot undo the damage if we decide to change our minds. To track these introductions is to delve into the historical side of "natural history" and to discover how interwoven this tale is with our own.

Each person, town, or country has creation stories, whether a legend of gods born from chaos, a child's cherished tale of how her parents met, or the story every fourth grader in southwestern Montana knows about how glacial Lake Missoula filled the valley before the ice dam broke for the last time and sent millions of gallons of water rushing into the Columbia River and out to sea. So here is another creation story for our country, already gifted with so many.

The plot centers on a ragged band of heroes, adventurers, and villains who have reshaped the United States because somehow we as a species wanted it that way. We chose starlings and gypsy moths and honeybees, just as clearly as we chose the Grand Coulee Dam and the Sears Tower. What we have ended up with after centuries of experimentation is an ecosystem at risk, biodiversity in decline, and a scramble to eradicate exotics and reintroduce natives. But as weevils lay eggs in the knapweed roots, mountain goats pick their way over Olympic peaks, and house sparrows flit through the alleys, we are witnessing the manifestations of all our desires.

1

INITIAL

FORAYS

CHAPTER ONE

THE PIGEON'S PROGRESS

IN a parking lot under a four-lane overpass, pigeons pluck bits of popcorn and hot dog bun from among the cigarette butts and broken glass. Some of the birds are built along classic lines, close to the ancestral rock dove: a slate gray body, dark bands on the wings, a white patch above the tail, an iridescent green and purple sheen on the neck. Except for hot pink feet and orange eyes, they are the colors of asphalt and shadow-striped cement. Others seem rusted, reddish brown in patches, or mud-splattered, dripped with black or beige. Another is gray with wings half white, as if it had soared through a narrow corridor, slick with wet paint. One, all white with bright black eyes, looks as though it could have borne holy messages before it started slumming down here just off the curb.

In a way, these feathers are a map back to each bird's origins, though we don't have a complete key. Scientists say the ones nearest to the wild form, called blue bars, use their exceptional skills as fliers to soar from the city deep into the countryside in search of grain. Aggressive, they defend larger territories than do their mottled cousins but are more susceptible to toxic poisoning. The explosion at Chernobyl killed more blue bars than any other type. Splotched, streaked, and multicolored birds fare well over the winter and breed year-round, signs of their links to a domesticated past, while white pigeons, rare and highly urban, lose aggressive contests, fail to defend wide areas, and make easy targets for predators but may be able to soak up toxins like no others.

The patchy tails and splotched wings also provide a different kind of history, the history of a population once domestic, that has been living wild for hundreds of years. Traits bred for over centuries are mixed and remixed under these steel girders and others like them throughout the United States, pairing speed with navigational skills, shiny feathers with sweet flesh. The wing tips—black, gray, brown, and white—all point back to one day in 1605 on the Canadian coast as a ship with a damaged rudder headed east.

ᚼ   ᚼ

THEY had failed again. New France, it seemed, would never take root on this sandy soil. In 1604 the sieur de Monts and his navigator, Samuel de Champlain, had set off to North America with high hopes, a monopoly on the fur trade, and a boatful of colonists. In exchange for his royally granted monopoly Monts must only establish a village and import fifty settlers a year. He had visions of heaps of thick beaver fur piled at the docks anticipating his ships, while Champlain thirsted after that long, rugged Atlantic coastline, his to map and explore. Any one of those rushing river mouths might lead to China and the wealth of silk and spices waiting there. As they followed their compass to the unknown continent, they plotted how to sow and tend the seeds of a glorious New France.

But nothing worked out as planned. The first colony at St. Croix, a small island at the mouth of the St. Croix River between present-day Maine and Canada, was a disaster. The settlers spent the summer fighting back swarms of mosquitoes as they struggled to build their new houses. They planted and watered gardens, only to watch the sun dry the soil, wilting young shoots. Legendary rich copper mines remained legends. Over the course of a long, brutal winter, man after man came down with scurvy. Their limbs swelled like balloons; their teeth grew so loose they could be pulled out by hand; eventually they died. As the survivors sat melting snow for water and heard the wind beating about their wood shelters, France must have lingered as the warmest of memories. Before the next cold season they abandoned St. Croix and built homes and planted crops again at Port Royal, tucked in a bay on the inland side of what became Nova Scotia, where the weather might prove milder.

Low on morale and supplies, they deserted Port Royal too after a year. Though his matter-of-fact journals chart no regret, Champlain must have felt bitter as he left the garden pools he'd filled with trout and watched those unknown, unnamed swells of land behind them disappear as the ship pulled away.

In a stormy sea, after breaking, then patching their rudder, the colonists heading back to France saw a small ship coursing toward them. Once within earshot, a man on board hollered over. The *Jonas*, a bigger ship, was sailing even now to Port Royal, laden with new supplies, he said. Reversing their direction, the settlers returned to their abandoned plots on the shore. The larger ship had beaten them home. Cannon blasted in celebration, the colony was reclaimed, a recently arrived nobleman opened a cask of wine, and they drank.

Along with hope and alcohol, the *Jonas* carried pigeons. While North America was home to several species of pigeons, including wood pigeons and, more notably, passenger pigeons that nested in clusters miles wide and blotted out the sun when a flock flew past, the domesticated rock dove, *Columba livia*, was a stranger. This ancestor of the street bird that now dirties statues in New York City and roosts on window ledges in Minneapolis originally inhabited a swath of land from the British Isles to India but had been bred and transported for thousands of years for food and amusement.

Champlain didn't mention the pigeons in his extensive

records, but then, he didn't mention his wife either. Focused on mapping a New France, Champlain spent his ink and enthusiasm on details of rivers and soil. But Marc Lescarbot, a passenger on the *Jonas*, a lawyer, and a sometime poet, did notice the pigeons and included them in his detailed notes about life at Port Royal. After describing the settlement's sole sheep in his *History of New France*, he writes, "We had no other domestic animals save hens and pigeons," and adds that the settlers had to watch out for eagles, which feasted on the plump birds as an imported delicacy.

Lescarbot didn't stop with pigeons. He chronicled a New France bursting with life overlooked by his gold- and fur-seeking peers: lobsters, mussels, and cod; thick forests and raspberries; eagles so plentiful as to be a nuisance. He also recorded the first of the flood of European species that colonists would bring to New France, establishing a patch of Europe amid the New World wilderness: wheat, rye, hemp, asparagus, French beans, turnips, peas, roses, hollyhocks, tulips, sheep, cattle, pigs, horses, geese, ducks, and donkeys. Another exotic—less desired—crept on board as well. Detailing the local tribes' encounters with European rats, Lescarbot comments, "The savages had no knowledge of these animals before our coming; but in our time they have been beset by them, since from our fort they went even to their lodges, a distance of over four hundred paces, to eat or suck their fish oils."

A year after Lescarbot left, the members of the settlers' party offered further evidence of how central pigeons were to their design. After abandoning Port Royal a second time, Champlain stubbornly returned yet again to North America, this time to forge down the St. Lawrence River and found Quebec. He sketched a drawing of the settlement, labeling each area meticulously. The picture shows a building half fanciful and half martial, part country village and part fortress. Three main sections housed quarters for workmen and artisans, a storehouse for weapons, a forge, and a kitchen. Raised outdoor walkways and a lower promenade allowed residents to stroll without leaving the main structure, and a drawbridge by turns permitted and forbade entrance. The sketch captures beefy curves of smoke curling out of three chimneys and similarly cheery swirls pouring out of three cannons. A flag above a sundial seems to snap in the breeze. A plot decorated with intricate designs indicates Champlain's gardens on the banks of the river. Towering three stories high, in the middle of it all, was the pigeon house.

Perhaps a pair of rock doves escaped and made a nest of sticks and feathers on a bare cliff and braved the snow rather than return to the elaborate pigeon house. Perhaps the pair spent the spring stealing hemp seeds out of Champlain's gardens and raising chicks and finally engendered a flock that soared gray over the town of Quebec, grew with it, and then began to push west and south. In any case, plenty more arrived with settlers from England flooding into Virginia. An early governor of Massachusetts received some as a gift. Pigeon houses rose in New Orleans. Cage after cage came over. Some odd and overwhelming desire created ample opportunity.

୬୧ ୬୧

DURING the years when Champlain mapped North America's east coast, pigeons represented nobility in its most refined and oppressive form. One of the rights of nobles in sixteenth- and seventeenth-century France, along with building windmills, taxing peasants, and administering a personal form of justice, was to own a dovecote, or pigeon house. Surrounding farmers were not allowed to kill the birds, no matter how many crops the pigeons devoured. At times more than a million flew over the fields, skimming choice grain. Early in the French Revolution the peasants released the nobles' pigeons, sending their privilege fluttering to the winds.*

Maybe in importing pigeons, the settlers of New France were trying to carry the structures of class with them, to set up the old society on new land across the sea. Unlike many colonists from England, who sought to escape oppression, French colonists wanted to spread their nation's political, religious, and biological glory. Though not themselves missionaries, both Champlain and Lescarbot lauded the efforts of Catholic priests to enlighten the Indians. While the missionaries worked to impart French morals to the natives, the settlers labored to give the land French seeds, both literal and figurative. The symbol of France is three lilies, and patriotic metaphor often painted France as a fertile garden. Champlain followed this blend of biology and symbolism when he wrote to the queen of his love of navigation, which "induced me to expose myself almost all my life to the impetuous waves of the ocean, and led me to explore the coasts of a part of America, especially of New France, where I have always desired to see the Lily flourish." The birds served as a similar sign of national pride, a feathered badge of patriotism, a reminder of French culture, class structure and all. As the bona fide noble of Champlain's party, Monts would have had the right to set up a dovecote wherever he claimed new land.

When sailors first jostled the pigeon coops from ship to land, the birds might have given gasping alarm calls at the cannon shots or struck out with wings to defend their nests. But as the cele-

---

*In some cases, the pigeon house was taken as a symbol of the whole regime. A visitor to France one hundred years after the French Revolution found a farmer living in a former noble's fancy house. He commented, "One relic of the good old times, still preserved in perfect repair in the gardens, tells how absolutely necessary were the great changes of 1792. It is a colombier, a pigeon house of gigantic dimensions, as large in fact as a church tower, which would accommodate some thousands of pigeons, which were allowed to devour the crops of the poor tenants in order to garnish the table of the seigneur."

bration wound down and the colonists went back to plowing and building, the pigeons probably started to settle in, nodding heads and preening, pecking at their grain, "coo-roo-cooing" and calling one another to new nest sites in this unknown country. The throaty coos of the imported birds reached the ears of all the settlers, commoners as well as nobles. Even if they resented the pigeons as aristocratic, the carpenter who measured posts for cannon platforms and the stonemason who listened to the churning of Canada's strange tide might have liked them as reminders of home. Amid all these new hawks and fish and seals and moose, what could be more of a comfort than the puff-breasted soft gray bodies bobbing in that homely strut?

Wherever it lands, the pigeon inspires a tangle of emotions. The rock dove, *Columba livia*, is one of the only animals with two vernacular names, one bearing all the complaints, the other holding all the praise. Few people look at the grimy-feathered, orange-eyed, randy birds defacing a statue in San Francisco's Union Square and think, "These street doves are a menace," just as hardly anyone hails "the white pigeon of peace." "Pigeon" originally referred to any young bird and comes from the Latin *pipire*, to chirp or cheep. The earliest listings appearing in the *Oxford English Dictionary* are for recipes ("Take peions and stop hem with garlec ypylled and with gode erbes," recommends a 1390s tome), but it's all downhill after that. "A pigeon" is a person who's easily duped or a coward; "to pigeon someone" is to take advantage of his or her gullible nature. A "stool pigeon" is nobody's friend.

"Dove," of more doubtful origins, may spring from the Old English word for "dive," but it has come to represent our hopes for our best selves. Unlike the 1390s quotation containing "pigeon," a use of "dove" from the same time period is not a recipe but a revelation. Theologian John Wyclif wrote: "The Spirit cam doun . . . and his Spirit was his dowfe [dove]." Doves decorate statues of early fertility goddesses, and a dove that returned with an olive leaf indicated receding water to Noah and his seasick ark. In Christian tradition, a dove is the holy spirit made manifest. We use the word to describe peace-loving politicians and to flatter those who make our hearts thrill ("She is coming, my dove, my dear," Tennyson exults in his poem "Maud"). Essentially the colonists brought doves to the New World and ended up surrounded by pigeons.

⁂

FOR a long time, in areas settled by both French and English, both before and after the French Revolution, dovecotes afforded succulent fowl for the dinner table. But agricultural changes pushed traditional dovecotes out of fashion during the nineteenth century. Meanwhile pigeon breeding picked up as a hobby for gentlemen. Birds were bred for plumage that curled out rather than in, short beaks, long legs, necks without feathers, and good posture. Many varieties looked like ordinary pigeons viewed in a fun house mirror. Tumblers flipped over backward in the air in a flurry of feathers, an excellent parlor trick. Pouters' chests swelled so high that breast feathers often buried their faces. Fantails displayed prominent tail feathers bristling in an erect semicircle, like a peacock's. Runts, bred for size, boasted wingspans of three feet. Today the most famous of the pigeon fanciers is Charles Darwin, whose observations of the birds formed a foundation block of his argument for natural selection in *On the Origin of Species*. He watched and took notes as offspring of white and black pigeons produced slate blue chicks with black bands on their wings, mirrors of what he believed was the ancestral form. He gently ridiculed his fellow enthusiasts who believed prehistoric England to have been populated by tumblers and fantails and pouters and runts. He proposed instead that people had selected for and enhanced these traits in the humble rock dove. And, if artificial selection worked, he reasoned, why not natural selection?

The feral pigeons that strut across the fire escapes have had both artificial and natural selection on their side. Survival of the fittest and survival of the most pleasing have sculpted a remarkable bird. Large eyes give pigeons a visual field of more than three hundred degrees, and they are able to detect near-ultraviolet light. Their sense of hearing, more acute than ours, allows them to track low and distant noise. Unlike most other birds, they feed their young on crop milk, a protein-rich fluid secreted from sacs in their throats. In addition, one of pigeons' most outstanding traits is their ability to breed and breed and breed. Though laying only one or two eggs per clutch, feral pigeons mate and rear chicks all year round in cheap nests composed of a few gathered sticks. These escaped domestics are more fertile than the wild rock doves still perched on the cliffs of Scotland, indicating that over the years humans selected good layers.

Another skill that rock doves have developed both naturally and artificially is navigation. Paired with an uncanny ability to recognize the spot where they were raised, navigational skills let them find their way home from hundreds of miles away. From the time of Caesar, armies bred pigeons to carry vital messages, but interest in pigeon navigation surged in the 1870s. After the Franco-Prussian War, when pigeon couriers distinguished themselves in the siege of Paris, both America and Europe were struck with what one observer called columbophilism, or pigeon love. Racing pigeons and homing pigeons, called homers, inspired particular passion. Fanciers rode newly developed railroad lines far from home, set their prize birds free, and claimed victory for the first one to make it back. Navy yards on both coasts built messenger pigeon stations. Newspaper reporters filed their scoops, and vote counters sent election returns via pigeon. In *Harper's Weekly*, Americans read about efforts of the New Zealand postal service to establish mail delivery by pigeon between two islands, complete with special "pigeongram" stamps, one shilling each.

Even as communication systems improved, homing pigeons

lingered. During World Wars I and II, Cher Ami, Lady Astor, Wisconsin Boy, and Jungle Joe all flew vital missions, and the public lauded the birds as heroes and awarded them medals. G.I. Joe, an American-bred homer, flew to a British fighter plane squad in time to stop it from bombing an area containing its own soldiers. Crippled by shrapnel and lamed by bullets, such birds were honored, then stuffed upon death. Their determination was credited to courage rather than instinct, as the praise offered by one pigeon lover suggests: "Into the breach went the little racing pigeon—the most gallant little bird the world knows. And they came through—came through with the messages of weal and woe; came through when the shattered troops were crying for aid—when every other line of communication had failed." It's hard to believe this is the same bird pecking at Doritos crumbs in a scum-covered puddle by the subway stairs.

All this tending and breeding gave rock doves a final talent, crucial to their success in large cities around the world: the ability to coexist with humans. Sadly, other members of the order containing the rock dove failed to breed themselves into abundance when faced with buildings and bullets. The last dodo, a huge flightless pigeon, became extinct on the island of Mauritius the same century that rock doves first soared above the New World. Just as urban rock dove populations began to explode, Martha, the last passenger pigeon of billions in America, died old and infertile in the Cincinnati Zoological Gardens in 1914.

⋊   ⋊

THESE highly trained birds, gloated over, fed choice tidbits, and awarded ribbons and medals, often never made it back to their lofts. A signal missed, a direction misjudged, and they were lost. Thousands joined birds that strayed from their rural dovecotes and rarefied breeds that escaped from pigeon fanciers, forming

pairs, then flocks, claiming window ledges, church towers, and city streets.

Only in the twentieth century did pigeons launch their takeover of American cities. In 1883 an article titled "Bird Life in the Central Park" curses the flocks of noisy sparrows but doesn't mention a single pigeon. But by 1915 naturalist Charles Townsend had watched hundreds flock to the Boston Common as tender-hearted souls scattered grain and the city fathers contemplated pigeonproof architecture. He noted that while the bird adapted to the bustle in the city, it had not yet figured out cars and "not infrequently lingers too long in the street and is run over or hit by the automobile while it is attempting to fly away." As late as 1939 San Francisco felt moved to install birdbaths to attract wildlife to Union Square, a decision it quickly regretted as the gray hordes descended. Some dramatic increases are even more recent: Between 1972 and 1992 the number of feral pigeons in El Paso, Texas, increased from roughly one hundred birds to more than fifteen hundred. Cracker crumbs and leftover lunches offered a feast for any creature able to tolerate a habitat of metal and rock. The pigeons not only were able to tolerate it but thrived, as if just waiting for us to build the environment of their dreams.

Abundance fosters distaste. While some successful exotics will drive other species close to extinction, grabbing their food, claiming their nest sites, passing on disease, pigeons mainly pester us. The birds swarm in parks, sometimes hopping on one leg or brandishing wings missing feathers, grisly casualties of urban life. They gobble seedlings planted by park groundskeepers. Pigeon droppings plaster the ledges of expensive office buildings. Dirty feathers host ticks and mites. Piles of dung can act as a breeding ground for a fungus that causes histoplasmosis infections in humans, particularly in those paid to clean them up.

The birds deemed "pigeons" (certainly not "doves" by this point) also mirror us in ways we may not find flattering. They live in cement-bound cities when they could fly anywhere. Once the pets of aristocrats, they now share the habitats of homeless men and women trying to get out of the rain. They achieve huge densities and cover their immediate surroundings with grime. They become greasy with dirt and filth. They're mostly monogamous but not strictly, and they prefer Twinkies and Wonder bread to more wholesome foods. But that's when we think about them. Mostly we don't.

For the stockbroker or custodian rushing to work, pigeons become as invisible as parking meters and chipped pavement squares, obscured by neon signs and store window mannequins posing in summer suits. The birds almost cease to seem like natural creatures, becoming just another city prop. But they live their whole lives on display. On a shelf of a bridge support, a white pigeon and a blue bar alternate brooding a clutch of eggs. On a windowsill bristling with antipigeon spikes, a checkered bird pecks at two fledglings, tapping them on the head and neck, urging them to test their wings. One approaches the ledge and pulls back, approaches and steps back again, until finally it spreads its feathers, braces itself, and flutters four feet to the base of a cement column.

Back under the overpass the multicolored pigeons feed, heads bobbing as they peck at the dirt. The males puff out their chests and fan their tails, trailing the females like gaudy shadows. Their cooing blends with the *shussh* of tires above. If a female doesn't chase him off, a male makes preening gestures around her head and neck. Then she pecks at his bill as a nestling asks for food, testing the male's paternal instincts. If he feeds her, the mating and nest building begin. It's a lifelong pair-bond that can produce chick after chick, but this particular romance is interrupted. A man who's been sleeping in a taxi emerges from deep in his beard and starts the engine. They're off, wings clapping.

Slowly even these feral pigeons that we loathe or dismiss may be shifting again in the public eye, working their way back into our good graces. In San Francisco, which for a long time fought back with poisons, spikes, and pigeon contraception, glossy posters appeared several years ago on kiosks and in bus stations offering commuters tips on pigeons' habits and natural history to heighten appreciation. The same year the city released white doves at a celebration for the newspaper columnist Herb Caen. Maybe in time we'll just accept these messy neighbors, yield up our statues and park benches as tribute, and let them go.

꙰   ꙰

CHAMPLAIN was a navigator and a geographer, adept at finding his way from place to place. Proud of his profession, he wrote books for seamen, instructing them to remain sober and become intimate with their astrolabes. He also wrote to the queen: "Among all the most useful and admirable arts, that of navigation has always seemed to me to hold the first place; for the more hazardous it is and the more attended by innumerable dangers and shipwrecks, so much the more is it esteemed and exalted above all others, being in no way suited to those who lack courage and resolution." His maps bear the marks of this ardent desire and seriousness of purpose. A sea beast with the head of a lion spouts water and frolics off the coast of Greenland. A bear and a moose as big as mountains roam through valleys. Bushy trees speckle the interior. Out at sea, two compass roses offer directions, along with sketchy longitude and latitude lines. While these imaginative flourishes fill the blank page, they reveal how little rather than how much was known. The new country in all its openness took root in Champlain as much as France took root in the new country. In a way he was the carrier pigeon of his day, sails like wings whisking him back and forth across the Atlantic, ferrying messages, seeds, animals, and ideas from shore to shore.

But while Champlain's primitive instruments directed him

over the salt waves, the pigeons, which tell time by the sun with eyes that can watch it move, which feel the tug of the earth's magnetism and lay out a map, which may smell dust from their loft on wind currents or hear distant waves to guide their way, didn't find their way back to France. They stayed and flourished in ways New France's first colonies never did. They remained as the British took New France. They dispersed as the French staged a revolution. They flew from Atlantic to Pacific as the Americans took the aftermath of their own revolution and built a nation with liberty, justice, and pigeons for all.

CHAPTER TWO

THE LAND OF MILK AND HONEY

THE swarm was cast.

First one, then five, then a hundred bees streamed from the old hive. The colony took flight, shattering into thousands of buzzing individuals for an instant, filling the air with their humming, before congealing into a mass on a tree limb near by. The way station achieved, they paused for instructions. Wax glands filled, gorged on honey, the bees basked in the sunshine of a warm afternoon in spring of 1622, ready to start building at a moment's notice. The queen too dragged her swollen abdomen from the old hive to this bare and promising branch where the swarm hung like a oversize pear, waiting for the sign.

Meanwhile the scout bees went exploring. They took off in all directions from the mass and flew a zigzag pattern, back and forth, pacing hollow trees and rock crevices, walking the insides and inspecting the outsides from all angles. They noted if the cavity could hold a winter's supply of honey, if the entrance faced south for warmth, if the site was high enough to thwart predators. Returning to the humming branch, the scout danced the distance and direction of her find, waggling vigorously if the site was excellent, shaking listlessly if it was only satisfactory. When a scout watched another wriggle with enthusiasm, she zoomed off to investigate for herself. If she approved, she returned and danced the new dance, spreading the news from scout to scout until all the scouts danced the same dance, waggled the same story on their hive mates' backs. Then they gave the signal for the swarm to move into its new home, and all the bees took flight again. While North America's native bees, unaware that competitors had arrived, went about their business of gathering pollen, the first colony of domestic honeybees to reach Virginia, to reach the New World, had split in two.

If a Jamestown man had looked up from his attempts to coax crops out of the swamp and glimpsed the gray cloud of the swarm gathered like fury, he might have interpreted it as an evil omen. Or smiled at it as a divine gift and anticipated honey the coming winter. Or he might have shrunk in disgust from these insects reputed to breed from the dead. What the legends didn't say, as this battered colony of England teetered on the verge of failure, was that in the pollen baskets on the back of their legs, the bees carried the grains of success.

≈   ≈

BACK in London, four months earlier, in late 1621, the stock-holders of the Virginia Company grew impatient. Hadn't they bled money as if from a knife in the side for more than ten years now? Yet with every letter from Virginia came more complaints. Please send more food and fewer people, the colonists wrote. We don't have time to plant and harvest enough for ourselves, and more hungry settlers arrive on each ship, expecting to be fed. They enclosed long lists of items each newcomer should bring: two bushels of peas, two bushels of oatmeal, a suit of light armor, three pairs of Irish stockings, a kettle, a grindstone, a gun. The investors replied by urging the colonists to work harder. Where are our profits? they asked. Where are the silk and wine and iron, the fruit of our seed money? We are sending one hundred more colonists to speed up production. The letters dashed back and forth, as fast as the ships could carry them, penned in tones of polite outrage.

In late November and early December 1621 the Virginia Company sent the *Bona Nova*, the *Hopewell*, and the fur trader *Discovery* to Jamestown. Over the course of four months the ships traveled between the fantasy of the adventurers and the reality of the suffering colonists, a gap so vast that the actual mileage seemed small in comparison.

The colonists' disillusionment and the Virginia Company's expectations were fed by report after report from explorers who described the New World as a paradise. In 1609 a Virginia Company ship named *Sea Adventure* disappeared on its way to Jamestown. Sir George Somers and 150 others were given up for dead until a year later, when two boats came to shore at Jamestown bearing crew members and passengers who had lived on the Bermuda islands for a year. From their tales of the easy life and plentiful food on the islands, Captain John Smith commented on "what a paradise this [was] to inhabit." Indeed these accounts of the Bermudas may have inspired Shakespeare to write *The Tempest* in 1611. In his version the island featured airy sprites, young lovers, and goddesses conjured out of the air.

Virginia too was painted with impossibly glowing colors. In a report on the state of the Virginia colony, Edward Waterhouse declared it "a Country which nothing but ignoarance can thinke ill of, and which no man but of a corrupt minde & ill purpose can defame. . . . It paralelleth the most opulent and rich Kingdomes of the world." He also added that cows grew bigger in Virginia, horses were more beautiful and courageous, and deer were so fertile that does dropped three fawns at a time. Smith, who didn't shy from relating the difficulties and deaths many colonists found in Virginia when he wasn't busy spreading tales of his own heroism and rescue by Pocahontas, lauded its possibilities: "These waters wash from the rocks such glistening tinctures that the ground in some places seemeth as guilded, where both the rocks and earth are so splendid to behold, that better judgements than ours might have been perswaded, they contained more then probabilities." Another visitor described "the purling Springs and wanton Rivers everywhere kissing the happy soyle into a perpetuall verture, into an unwearied fertility." As scouts, dancing wildly, they made up in enthusiasm for what they lacked in accuracy.

While these visions sparkled before the eyes of the stockholders, colonist Peter Arondelle wrote to England in hope of

gaining some relief from the grim reality hovering behind them. In a letter to Edwin Sandys dated December 15, 1621, just after the ships left England, he wrote: "these few private lines shall only serve to intreate your favorable voyce vnto the Company, for the pformance of their promises. And because I am neerer to me than any other, and that Charitie begins wth ones self, I crave pticularly for me and my poore familie. whereas Mr Deputie Ferrers promised me the assistance of Captaine Nuse and my Sonne in lawe Captaine Mansell (who is dead) for fishing and hunting, and provision for a whole yeare before hande, a house ready builte, and Cattell: wch proved farre defectyve. For, for provision all that wee now have is but a pinte and a halfe of musty meale for a man a day."

But relief would be a long time coming. Malaria-bearing mosquitoes continued to rise out of the swamps and kill off new arrivals. Eager for a crop with value in English currency, the colonists planted tobacco rather than corn, forcing Jamestown to rely on shipments from England and trade with the Native Americans rather than on self-sufficiency. Every year brought threats of starvation, and the colonists had to cajole and threaten the Powhatan Indians to provide them with supplies. Houses proved so scarce that one visitor reported that newcomers had to seek shelter under bushes. Moreover, tensions with the Native Americans were building.

Part of the colonists' disorientation arose from the fact that this new land was so unfamiliar on a biological level. The Native Americans around Jamestown farmed, but their crops—squash, corn, kidney beans, and other foods—were unfamiliar to European farmers. To supplement their harvest, the Native Americans hunted, fished, and sought edible leaves and roots with intimate knowledge of the lives of the animals and plants they sought. To really settle into this new country, rather than perch on its edge and try to skim money from it as quickly as possible, the colonists required much more than houses or fences. They needed either to observe and adopt Native American ways of using indigenous flora and fauna or to import a biological infrastructure of trees and flowers, birds and insects whose ways they knew.

Early on the morning of March 22, 1622, four months after the Virginia Company and Arondelle sent their letters and a few weeks before the boats were to reach Jamestown, the Powhatans attacked. Perhaps they had grown concerned as more and more settlers had come in with each boat, wanting land, demanding corn. At breakfast, in the fields, indoors and out, they picked up weapons and cut the settlers down, taking advantage of the fact that the colonists had moved out from beyond the protective core of the original town. The slaughter astonished the colonists, who could not see how the natives would ignore their offers of Christianity and their restraint in not claiming their prime fields.

As the survivors tallied their losses, the establishment of Jamestown seemed all the more precarious. The men sent to work iron, all killed. Houses burned. Small starts of vineyards torn by the roots. They vowed revenge and determined to crush the natives in whatever way they could but wondered if they had the means. By the end of 1621 the Virginia Company had sent over 4,270 colonists. Just before the massacre 1,240 remained; in its wake there were a mere 893 settlers left. To make this a paradise— the land of milk and honey—they had a lot of work to do.

     ⁂

A FEW weeks later, in mid-April, as the colonists were still sorting through the rubble of their homes to find usable scraps, the *Discovery*, the *Bona Nova*, and the *Hopewell* blew into port. As the remaining residents unloaded the ships, the governor and council in Virginia opened a letter from the company in London that urged increased productivity and listed the contents of the holds:

"We have by this Shipp and the Discoverie sent you divrs sorts of seeds, and fruit trees, and also pigeons, connies, peacocks maistives and beehives. . . . We have sent unto you likewise some vine cuttings and a very smale quantite of silkworm seed." They also sent twenty new colonists in the *Discovery*, twenty in the *Hopewell*, and fifty in the *Bona Nova*.

Most items of cargo on the ship had clear and well-described functions. Seeds and fruit trees hinted that the colonists should establish gardens in addition to tobacco fields. At the same time King James I harbored hope of using the colony to break England's dependence on imported silk, and he personally contributed several batches of "silkworm seed" (cocoons) to the effort. The Virginia Company instructed the governor that no colonists except council members and heads of plantations would be allowed to wear silk unless they produced it on Virginia soil. Wine too held promise. Explorers discovered grapes growing wild in the New World, so winemakers were imported along with vine cuttings from France. Pigeons and rabbits (coneys) made tasty meals, if they were a bit redundant with the abundance of wild game. The bees fitted well with these items. Not exactly necessary, like a pair of Irish stockings or a bushel of peas, but they would sweeten a bitter winter and add gloss to life as did silk or wine.

Honey and wax made bees an essential part of 1600s agriculture. In 1623, one year after the first honeybees arrived on Virginian soil, Gervase Markham wrote *Farewell to Husbandry*, a how-to book for tilling the land. Among the many tasks that a farmer was supposed to accomplish during the month of April was to "open . . . hives and give Bees free liberty, leave to succour them with food, and let them labour for their living." While farmers valued bees because they offered honey to preserve food and wax to make candles, they also saw them as a way to make money through trade. According to E.W. Gent in "Virginia: More Especially the South Part of, Richly and Truly Valued," both honey and wax made the list of Virginia exports in the 1620s. Mid-century a planter named George Pelton earned thirty pounds a year off his hives, and potential stretched far beyond this sum. The author of "A New Description of Virginia" reported that "if men would endeavor to increase this kind of creature, there would be here in a short time abundance of Wax and Honey, for there is all over the country delicate food for Bees."

But the interest in honeybees, or *Apis mellifera*, was more than economic. The same mixture of science and fantasy that allowed explorers to categorize the natural resources of Virginia with a practical eye, while partially believing they stood at the doorstep of Eden, allowed them to look at animals simultaneously through the lenses of money, morality, and magic. In some ways the organization, hygiene, division of labor, and selflessness of honeybees all seemed like an example for humans. Some writers even considered these six-legged workers and drones better citizens than men, who could be disorganized, dirty, chaotic, and selfish. Naturalists looked to Pliny as their greatest authority. He had lauded the bee: "Nature is so great that from a tiny, ghost-like creature she has made something incomparable. What sinews or muscles can we compare with the enormous efficiency and industry shown by bees? What men, in heaven's name, can we set alongside these insects which are superior to men when it comes to reasoning?" Virgil, coming back in vogue in the seventeenth century, had devoted the fourth section of his agricultural poem *The Georgics* to bees, crediting them with all kinds of virtues: "I'll tell of a tiny/Republic that makes a show well worth your admiration—/Great-hearted leaders, a whole nation whose work is planned." Montaigne, not long before the Virginia Company staked its claim, had also praised the honeybees for their attacks on other bees and for their use in human combat. He cited battles in which one side would fling bees at the other to rout opposing forces: "[Bees] will have the power and courage to scatter an army." And the product of all this tidiness, organization, and aggression? Honey, sweet gold. Enough to make the hard months of winter easy. Enough, in the right dose, mixed into royal jelly at the right time, to turn a worker into a queen.

It's difficult to reach back and grasp what type of world the adventurers and colonists lived in, so different from our own. While imbuing the natural environment with moral qualities, they based their knowledge on scholarly reports rather than on the direct observation we consider vital to science. At the seashore they were told a special breed of geese grew perfectly formed in the center of barnacles. In the sky eagles carried cool stones to their nests to ease their labor pangs. Deep in the forests unicorns paced, always just out of sight, while trees dropped leaves that turned to swallows before they hit the ground. Underground ants foretold the future, and male and female diamonds mated to produce glittering offspring. The scraps of information that travelers and scientists gathered provided only fragments of the entire picture, and imagination rushed in to patch the holes.

The Jamestown colonists lived in a time perched on the edge of scientific breakthroughs about insects along with other facets of biology. In 1625 Francesco Stelluti peered at bees through a microscope, only newly in use. The fragmented eyes, the delicate veining of the wings, the hooked end of each foot: All came into clear focus for the first time. Stelluti recorded his observations in an engraving, with particular attention to the insect's long tongue. In 1669, as increasing numbers of honeybees plundered nectar from Virginia's wildflowers, Jan Swammerdam offered more accurate observations of bees than anyone to that time in his book *The General History of Insects* and suggested that spontaneous generation was a myth. By the end of the century Francesco Redi had established it without a doubt. Employing a strict regimen of observation and experiment, Redi rewrote the natural history books. Did swallows generate from leaves? No. Did fish in China sprout feathers in the summer? Highly unlikely. Did birds use magic stones to restore eyesight in their young? Definitely not. These revelations were right around the corner, but in 1622 ideas about the lives of insects were as fantasy-fueled as ideas about the New World.

One particular field about to burst into bloom with the help of the scientific method was the study of pollination. While botanists and gardeners knew that pollen affected plants' ability to bear fruit, they didn't understand the insect's role. In order to gauge its true importance, they had to learn to see bees in a new way. One hundred years after the first hive had arrived in Jamestown, scientist Philip Miller plucked the stamens out of all the flowers in one patch of his tulips to see if they would reproduce. As he watched, bees covered with what appeared to be dust flew from a plot of intact tulips to the ones without stamens, visited the blossoms, and left some of the powder behind when they buzzed away. Several years later Arthur Dobbs wrote in his *Philosophical Transactions* of 1750, "Now if the facts are so, and my observations true, I think that Providence has appointed the Bee to be very instrumental in promoting the increase of vegetables." The New World colonies would soon prove it.

Each spring, traveling from blossom to blossom, the honeybees gathered nectar to cure into honey and pollen to pack into bee bread. When, as often happened, grains of pollen caught on their fuzzy heads and bodies and brushed against the stigmas of each plant, pollen tubes crept down the stiles. The ovaries began to swell. Peaches, apples, oranges, avocados, and other plants benefiting from bees grew and sweetened, developing seeds, turning to juice, pushing against their tight skins, changing from cells of ovule and pollen to heavy fruit, fragrant and ripe. In the honeybees' wake, pears, cucumbers, and watermelon bloomed and bloomed and bloomed.

In the years following the massacre, the Virginia Company faltered. By 1624 inability to turn a profit had caused investors to waver in their commitment. A royal investigation shut the company down and returned power to the king. But the honeybees remained a lasting legacy, traveling as far north as they could bear the cold, pushing on to the south and west. Colonists tended some hives, extracting the honey-filled combs to sweeten their meals. Other bee populations turned feral and built their networks of combs in hollow trees. As their numbers grew in the years that followed, they continued to split the hives, swarm, and establish new colonies, keeping pace with the settlers. In his "Notes on the State of Virginia," Thomas Jefferson reported that the Native Americans observed the bees hovering constantly around the settlers and dubbed them "the white man's fly." As the colonists sought to remake the land they found in a European image, fenced and divided into fields planted with exotic vegetables, fruit, and grain, the bees offered invaluable assistance. Many of the crops relied on insect pollination, and with the multiplying honeybees, bearing pollen from plant to plant, the harvest grew.

Watching their familiar trees, vines, and shrubs take root in foreign soil and convert this wild landscape to one that looked almost like England, the settlers must have thought it all a bit like magic. A bit like paradise.

⚘ ⚘

HERE at Jamestown, about 375 years later, midafternoon light pours through the trees. Jet skis and motorboats race up and down the James River, while a ferry makes its persistent way upstream. A flock of seagulls swoops behind, cawing for scraps. The sun glints off the water, making tourists rue sunglasses left in the car.

At the site of the old settlement, history is commemorated in familiar ways. Captain John Smith, on a high pedestal, looks out over the river, one hand clasping a Bible, the other resting on his sword. Pocahontas hides behind the church, and an old man, gently holding her hand, poses for a photograph. Stacks of brick mark foundations of old houses, and recorded messages try to

breathe life into the ruins. Near the Smith statue, yellow posts of
an archaeological dig indicate the borders of the original fort, part
of which extends now into the river. A man in an Elizabethan cos-
tume, complete with plume and white doublet, tells the crowd
stories about the rowdy old days. Glass cases in the visitors' center hold a model ship, a brooch, a coin. Walking through
Jamestown today is more like peering into a mirror of our present culture than glimpsing a window to the past. On this bright
afternoon it's hard to imagine Peter Arondelle eating his musty
"meale" and writing his desperate letter back to England.

But on a smaller scale a different history unfolds. A yellow
butterfly lifts from the graveyard onto the black metal fence. A
small white spider casts billows of silk into the wind. And while
the earth of the settlement is hard and dry, supporting lawns and
benches, the ground quickly drops into marsh not far inland. A
bridge to the parking lot spans black mud and cattails, shreds of
the earlier story. Near the tollbooth a honeybee extracts nectar
from a pink blossom. The question rises, not why was it brought
here, but why, in this heat, with these flowers, is there only one?

In the wake of the honeybees' initial success, other introductions followed. Farther north New England settlers brought bees
as well, and the first colonist to carry over a hive supposedly
received a plot of land as a reward. In 1780 Colonel Herrod introduced honeybees into Kentucky. But by that time the insects were
so knitted into culture and agriculture and commerce, people
could hardly remember if they were native or not.

In a paper read before the Philosophical Society in Philadelphia in 1783, Benjamin Smith Barton took the position that
domestic honeybees did not originate in the New World and followed up with a long and elegant argument that cited many other
species, taken for granted in America, from the clothes moth to
the dandelion, that were actually European imports. He scoffed
at those who argued that honeybees had always been here and
suggested that certain authors on the other side of the Atlantic

might posit that honeybees were native to America but took on the unruly character of their country. They scattered all over, refusing to form the neat and industrious societies expected of their European counterparts, and that explained why they were so difficult to find. But in perhaps the most valuable aspect of his investigations, he scoured descriptions recorded by early explorers and naturalists and uncovered many strange reports: bees that made honey but didn't sting; tiny fly-size bees that built bizarre-looking honeycombs; bees that swarmed underground

Like Jefferson, Barton pointed out that honeybees only hovered near settlements and that pioneers kept discovering areas where there weren't any. As a result, the introductions continued. In 1853 Christopher Sheldon brought a hive's worth to California (a plaque in front of Terminal C of the San Jose Municipal Airport commemorates the introduction). Italian, Carniolan, and Caucasian varieties of honeybee appeared in 1859, 1883, and 1905 respectively. In the 1950s fierce Africanized honeybees were imported to Brazil. They have since spread northward into the southwestern United States, worrying beekeepers as they interbreed with the more docile European varieties. By the 1980s imported honeybees were pollinating four-fifths of the commercial crops in the United States, performing services worth millions of dollars to farmers.

But recently the easy relationship of food plants and honeybees has shown that dependence can be dangerous. As long as observers have watched bees work in their hives, they have also watched the hives collapse from disease. Foulbrood, chalkbrood, sacbrood, acute bee paralysis, and cloudy wing virus all can invade a colony and silence its buzzing. In the past decade, tracheal mites and varroa mites, two of honeybees' old enemies, have caught up with them in the New World. Tracheal mites live their whole lives in the breathing apparatus of honeybees, weakening their hosts as they clog their airways and suck their blood. Varroa mites, appearing ten years ago, make their way into uncapped cells containing young honeybees and breed in the larvae. The adult bees crawl out, often crippled and diseased, while a new generation of mites disperses into the brood nest. Together these two parasites have wiped out most of the feral honeybee colonies in the United States and are killing off some commercial hives as well.

While some beekeepers are developing remedies and mite-resistant strains, others are turning their eyes to the native bees that remain, more than thirty-five hundred species in North America. Like the pioneers in pollination studies, they are once again learning to look at bees in a new way. Shucking off the popular notion of a bee as a black- and gold-striped creature that makes honey, they are focusing on the scientific definition, an insect that mines pollen. They are also seeing things that may be as eye-opening as Stelluti's first sight of a bee foot under a microscope: mustached mud bees making their homes in clay walls, turquoise-banded alkali bees laying eggs in the dirt, carpenter bees digging chambers into wood and constructing sawdust rooms for their offspring, cuckoo bees stealthily dropping eggs in other bees' nests, and sweat bees licking the salt fluid gathering behind ears and on the inside of knees. Digger bees, leaf-cutter bees, plasterer bees, and orchard mason bees move from flower to flower.

For all their diversity, though, they aren't as plentiful as they once were. The transformation of the landscape into a place hospitable for European agriculture and European honeybees has been hard on the North American species. Native bees, some dependent on specific wild plants, others living in the soil or in dead trees, have declined as honeybees have taken over and crops have pushed back grasslands and woods. Some soil dwellers have died off as farmers have plowed under their larvae. Those making chambers in dead wood have disappeared as forests have been cleared to make room for homes and fields. Pesticides have killed others. Most native bees are solitary creatures rather than social insects like honeybees, and as a result they have found it hard to compete with *Apis mellifera*, which can locate a good patch of nectar and pollen and waggle instructions to a thousand sisters so they can help pick the meadow clean. The native pollinators have struggled. Until recently honeybees have rushed in to take their places, thriving on disturbance, living easily next to humans, working well with agricultural plants, mining the flowers at a rate the natives couldn't match.

With honeybee colonies suffering, scientists and beekeepers are turning their energies to rebuilding native bee populations and, perhaps more challenging, training them to work with people. Alfalfa farmers carry chunks of dirt containing alkali bee larvae to their fields, laying them in "bee beds" of soil prepared just how they like it, hoping that when they mature, they will head for the alfalfa crops. Locked in greenhouses, bumblebees are proving capable of buzz pollinating hothouse tomatoes. Some of the natives are even more efficient pollinators than honeybees, if given the chance, but an industry based on insects that evolved in North America, though more ecologically justifiable than reliance on nonnatives, will be no easy task. One scientist taught blue orchard bees to live in holes he drilled for them, but it took ten years.

Whether the mites will really wipe out the honeybees and whether native pollinators will be able to take their place remain to be seen. Without doubt, however, the threat of the honeybee's disappearance has exposed our dependence on an insect that aided all the colonists' efforts from the moment the first scout flew out to evaluate the Virginia woods. As beekeepers struggle to domesticate native pollinators and honeybee populations continue to falter, the true value of the cargo carried on the *Discovery*, *Bona Nova*, and *Hopewell* only now becomes clear.

CHAPTER THREE

WAR STORIES

IN 1776 General William Howe arrived in America under orders to subdue the colonists who were in open rebellion. With him, he brought mercenaries from the areas that eventually joined others to form Germany: Hesse-Hanau, Hesse-Kassel, Ansbach-Bayreuth, and Anhalt-Zerbst. These professional soldiers brought with them bayonets, cannonballs, maps, tents, salted pork, dried peas, seeds in the dirt of their boots, microbes on their skin, bacteria in their stomachs, and possibly a small fly, which may have traveled in the straw the soldiers used for bedding or to pack their supplies. The Hessian fly, or *Mayetiola destructor*, would thrive in America long after the Hessians and British had been evicted, eventually becoming the nation's worst wheat pest.

The facts of this introduction are as small and pesky as flies themselves, which buzz in the ear but remain out of sight. During a war no one is paying much attention to bugs, particularly ones that busy themselves with plants rather than people. Studying insects is a leisure activity, but once they cross the line into agriculture, the hobby becomes business. The effects of the fly's hunger are visible everywhere, in weak wheat crops, in struggling farms, but the cause is hidden in a cloud of cannon smoke.

Only when the tide of battle turned, when the rebellion, which appeared on the brink of failure, was pressing toward victory, and when General Howe was recalled to England in disgrace, did anyone pay attention to the scale of the fly invasion. In

1778 farmers on Long Island noticed that in the same spots where the Hessians camped, some wheat plants were stunted. Stalks grew bumpy, thick, with dark green leaves. Many shriveled up over the winter. Others failed to put out the tillers that allow the plant to expand. In 1779 the problem worsened. Entire fields of wheat, labored over and borrowed for, collapsed and broke on the verge of ripening. Rows that should have been packed with green plants turned yellow and shrank, revealing patches of bare ground. The Americans took even more notice as county after county in New York succumbed to the blight, and their main concerns switched from fighting to writing letters to the newspapers about the wording of the Constitution and repairing their homes and fields.

Amid organizing their government and sifting through ads for worm powders, stagecoach rides, opium, hemp cordage, new printings of *The Principles of Latin Grammar*, and rewards for runaway slaves, the Americans glimpsed a new enemy on their shores. So small they were barely visible at times, the insects hovered over wheat fields or clung to the plants, waiting for a spell of warm, calm egg-laying weather. Those who got on their hands and knees and looked closely saw, in the warm sunlight of late summer, pupae emerging from a piece of straw and flies crawling out. Tiny, delicate, the small brown bodies perched on long zigzagged legs, looking like mosquitoes that feed on plant juices rather than blood. After summering in the stalk, the flies were ready to mate and lay one to three hundred eggs, cradling them in the veins of the wheat leaves. When these eggs hatched, the maggots crawled down the leaves to the stem, where they began to feed, causing the stalk to swell and weaken.

By 1786 the insects were attacking crops in New Jersey and Pennsylvania. Some farmers abandoned their wheat fields, plowed them under, and planted corn instead. As the fly continued its ravages, a partial solution revealed itself. After the insects had destroyed the crops of Isaac Underhill, a Long Island farmer

and miller, he took seed stored in his mill to plant another round. He had purchased the seed from a ship docked in New York, and when he planted it, he grew fine, strong stalks of yellow bearded wheat, while his neighbors watched another crop dry up and break. The flies lived in Underhill's fields too, tucking eggs along the stalk, but the wheat seemed to shrug them off. Underhill had pulled from his mill a resistant variety, a wheat that had evolved to persevere through the attacks of the fly. Still, this offered only a small obstacle in the insect's progress, a boulder tossed in the river's path. Great Britain, afraid of importing the pest, banned wheat from America. More crops succumbed every year. The fly was troublesome and threatened their homes, and the Americans named it after their old enemy, calling it the Hessian fly.

But the evidence that the Hessians actually introduced the Hessian fly was sketchy while the motivation to blame them was strong. The colonists despised them as mercenaries interfering in a personal feud, and their hatred didn't soften any as the soldiers looted their homes, despite strict orders from General Howe. As Bernhard Uhlendorf points out in *Revolution in America*, the Declaration of Independence itself devotes space to condemning the transport of "large armies of foreign mercenaries to complete the works of death, desolation, and tyranny already begun with circumstances of cruelty and perfidy scarcely paralleled in the most barbarous ages, and totally unworthy the head of a civilized nation." In his recollections of his days as a soldier in the Revolutionary War, Private Joseph Plumb Martin writes of the Hessians, with a plaintive tone, "[T]hey should have kept at home; we should . . . never have gone after to kill them in their own country." Maybe the pinpointing of the insects' origin in a spot where the Hessians camped had a political motivation. Pestilence makes good propaganda. No one with a German accent came forward to admit that clouds of newly hatched flies had hovered over his bed at night, ovipositors pulsing, before heading out into the American wheat fields. Loyalists responded with the accusation that the Hessian flies were somehow all George Washington's fault.

Ten years after General Howe and his troops landed on Long Island, the name Hessian fly was firmly established, even if the origin wasn't. In 1787 a citizen sent a letter to the *Pennsylvania Gazette*, writing: "I would beg leave to propose . . . that the Philadelphia Society appoint a committee to inspect a number of fields of grain, infested with the Hessian fly, to inform themselves of its history and progress. . . . This insect did not advance to my neighborhood, to be observed, until May 1786. It is now increasing to an amazing degree."

In 1787 a pamphlet appeared offering another suggestion about the origin of the Hessian fly. A writer using the pseudonym The Prophet Nathan declared the plague a judgment and the insects themselves "manifest tokens of the displeasure of God, against the heaven daring sins and abominations of this land." The Hessian fly represented a heavenly test, and many people were flunking. Farmers who lost their crops, instead of humbling themselves before God, railed against him. "They have fallen almost or quite into a frenzy, in execrations upon the Hessians," Nathan reported. Some even took the Lord's name in vain. Those who had supplies of wheat resistant to the Hessian fly sold them for the highest prices they could get, milking the last pennies from their neighbors. The worst presented ordinary wheat as fly-resistant, swindling their fellow farmers as well as robbing them. Wherever it came from, the fly brought out the worst in people.

Throughout the nineteenth century the debate stayed alive, the identity of the Hessian fly obviously a lingering concern. Sir Joseph Banks, England's leading botanist and president of the Royal Society of London, conducted an investigation and declared he couldn't find a Hessian fly in all of Europe, dampening the enthusiasm of those who liked to curse the Hessians. Then, in 1834, a professor plucked larvae from a wheat plant in Minorca, an island off the coast of Spain, and watched as they developed into Hessian flies. Reports of invasions cropped up throughout the southern Mediterranean. Those who lived in the region told the professor the insect had always been around. Asa Fitch, an entomologist making a name for himself in America, leaped on this as confirmation of the entire Hessian fly tale, fly-ridden straw used for bedding or packing, Howe's march into Flatbush, and all. In an 1845 paper Fitch fancifully suggested that since infested fields were marked by damaged, broken stalks, "Had a company of soldiers needed straw for package, no objections would have been made to their going into a field of this kind, and with a scythe, gathering what they required."

Finally, the Hessians had enough. Balthasar Wagner, an entomologist from Hesse, set out to put the record straight. The 1861 rebuttal included detailed movements of the Hessian troops, analysis of the fly's life-span and its ability to survive a four-month-long boat ride, and a few snide comments about Asa Fitch. In the first place, Wagner argued, an insect exactly like the Hessian fly had attacked German wheat crops in 1857, and no one could remember ever seeing it before. In the second place, when a colonel in upstate New York heard about all the fuss in Long Island, he mentioned he'd seen those same insects in his wheat fields before 1776, well in advance of the Hessians' landing. In the third place, did anyone really think the soldiers brought the same moldy straw they'd been sleeping on when they left Bremerhaven in late March with them to attack Long Island in August? Wagner thought the insect came from Asia originally, along with the cereals it fed on, and added that the pest probably arrived in North America with grain from France. Everyone should call it wheat destroyer instead of Hessian fly, he suggested.

The article was overlooked, and Wagner's patient explanations were buried. The fly brought by the hostile army was too useful a symbol of foreign invasion to let logic brush it away. For instance, the very next year an impassioned antislavery tract contained this, slightly embellished invasion story: "Three Hessian flies only were seen on the cabin-wall of a Dutch ship which approached an American wharf; now what field in the continent has not known the devastations of the Hessian fly? Six Norway rats swam ashore from another Dutch ship in our waters; now where is the cellar without them? Two hundred and forty years ago, twenty slaves were brought to Jamestown, Va, in Dutch ship No. 3;—now where can you go, from Bunker Hill to Sumter, without hearing the rattle of a slave's chain?" The metaphor was flexible enough for a broad range of uses.

Eventually an American scientist, familiar with Wagner, peeved that everyone insisted on calling the insect the Hessian fly, and dismayed that the United States Entomological Commission repeated the tale about the soldiers and the straw in their last bulletin, reiterated Wagner's arguments in the *Canadian Entomologist*. "Patriotic motives are the worst guides in scientific questions," H. A. Hagen complained. He spent most of the article repeating Wagner's points in stronger language, but his unpatriotic conclusion was his own and a shocker: "I consider, therefore, the Hessian fly to be an indigenous American insect and not imported by Hessian troops."

In a way it is. Though Hagen's suggestions that the fly had always existed in America and no one had ever noticed it before 1778 are not that convincing, there's another way to look at the question. Wheat itself is an exotic—that is, if a plant so cultivated can lay claim to one region or another. In his 1908 book *The Book of Wheat* Peter Tracy Dondlinger cites sources that say wheat originated in Mesopotamia but notes that tracing the plant is difficult as even wild patches are thought to have escaped from cultivation. Wheat arrived in America along with the Europeans in the sixteenth and seventeenth centuries. The Hessian fly has followed the wheat wherever it goes and inhabits most, if not all, of the places worldwide where the grain grows. Maybe the question of native versus exotic becomes moot here since the "natural" home of Hessian flies seems to be in the leaves and on the stem of the plant they evolved to eat.

The fly remains. It eventually made its way into every state with a wheat patch. Pesticides never were that successful, and farmers have taken to planting around the breeding cycle of the fly and plowing infested fields under. Another strategy rises from those early 1780s reports of wheat varieties that stood straight and healthy in their fields while plants keeled over in adjoining plots. The larvae died on the resistant plants, unable to feed for one reason or another, and even as the fly pushed into new states, farmers uncovered more types of wheat that could live through the attacks. China, Red May, Red Chaff, and Mediterranean were some early resistant varieties. In the late 1800s farmers recorded that Palestine, Polish, Common March, Diamond, and Egyptian Imported proved at least somewhat resistant. At the turn of the century Prosperity, Democrat, and Red Russian joined the list. Pawnee, Omaha, Redcoat, and Ben Hur were recognized as resistant in the late 1920s. But in time the few Hessian flies that could thrive on these varieties predominated, and the breeders had to come up with something new. The Hessian fly and wheat have been together long enough to engage in an intricate evolutionary tango, and when natural selection encourages a wheat strain impervious to the fly, natural selection builds a better fly. Currently there are forty wheat varieties resistant to the Hessian fly, and others are under development all the time, as researchers try to outrun the one true fact about the fly's introduction: The hostile invader and welcome guest come traveling hand in hand.

CHAPTER FOUR

FLIGHT OF THE MOSQUITO

WHEN HMS *Blonde* docked at Hawaii in 1825, it carried a heavy burden. The island's young king and queen, Kamehameha II and his bride, Kamamalu, had sailed to England two years before, seeking protection from King George for the Hawaiian Islands and anticipating the wonders of the civilization they'd seen reflected in nails, ships, and guns. Instead they found measles, another by-product of the great cities. Because they lived on a remote island, their bodies hadn't developed resistance to diseases that had brewed in Europe's water, air, and blood for centuries so that the English, French, and Spanish were braced against them on a cellular level. Kamehameha II and Kamamalu died two weeks after stepping off the ship, and the *Blonde*, under the command of George Anson Byron, the poet's cousin, brought their remains home to their subjects.

No one got what they expected, not the young king and queen traveling abroad, not the islanders waiting to welcome their rulers home, not the Europeans, some with visions of Hawaii as a primitive dreamland. Ever since Captain Cook landed there in 1778, a steady stream of whalers, sandalwood traders, missionaries, and diplomats followed in his wake, not dissuaded by the fact that Cook had been killed by natives on his last visit. In places, instead of lush tropical greenery, the *Blonde* crew found parched earth, stomped by cows and goats brought by earlier European visitors. The natives frequently struck travelers as dirty or greedy and often sick.

Still, those on the *Blonde* with modest expectations were charmed by the feathery coconut palms and breadfruit trees clustered on the beaches and the waterfalls draining into the sea. In the mornings Mauna Loa volcano glowed with pink light, otherworldly and strange. In the afternoons the visitors could watch natives scooting canoes through the curl of a wave or pick the wild strawberries that pushed through the lava. In the evenings Hawaii's bright birds winged through the forest, pausing to coax nectar from lobelias with beaks curved to fit the flowers' long throats. If all that wasn't enough to lull the sailors into believing they'd found a scrap of heaven in the Pacific, there was this: not one mosquito.

ↄ ↄ

SLAP.

In 1826 Hawaiians living near Lahaina in Maui heard an unfamiliar whine. High and almost hysterical, it ripped the twilight like an arrow. Heavy with sleep, they'd be slipping away when the hum began, first distant, then ringing inside the whorls of their ears. Shaking their heads or waving their fingers brought only a moment's quiet. Swinging an arm toward the sound, they might catch an insubstantial insect, all legs and wings. Crush it, and their palms would cup a black smear of thorax, head, antennae, and a brushstroke of blood. In the silence they drifted into unconsciousness. In the morning red and itchy bumps mottled their bodies.

They brought their questions to Gerrit Judd, a doctor from America, and showed him their swollen arms and legs. While his main duty was treating the missionaries who came to Hawaii to offer Christianity to the islanders, he also tended the natives, performing surgeries, treating them for syphilis, measles, influenza, and other diseases that slid into port with whalers and merchants. Indeed, by the time Dr. Judd arrived, the native Hawaiian population had dwindled to less than half of what it had been when Captain Cook first landed, dropping from 300,000 to little more than 130,000. For a people so wrecked by change, this new inconvenience could hardly have appeared significant. It was just one more inexplicable shift.

The missionaries, though they knew their tormentor's name, still suffered. Craving relief from the voracious newcomers, Laura Judd sewed mosquito nets for her family out of calico, which kept the bugs away but almost smothered the children. In a letter to the missionary board, Dr. Judd wrote, in the tone of a man near the brink, "O the mosquitoes! Do buy all the mosquitoe curtains necessary for the use of all who are bitten."

Dr. Judd and the missionaries tracked the newcomer to the *Wellington*, a merchant ship that had come from Mexico to the port at Lahaina, one of the main stops for whalers seeking to restock supplies. A rowdy village, Lahaina catered to the tides of sailors that flooded in, frantic with shore leave, demanding sex and alcohol and quickly, and then drained away again. According to shipping records, 138 whalers docked at Lahaina in 1826. Merchant and government ships swelled that total. Hosting both a mission and a constant stream of sailors, Lahaina was a microcosm of the tensions that wracked the island chain. The scraps of ground and the Hawaiian people themselves were wrestled over with the ferocity of two dogs at a rope, though the missionaries and sailors were not playing. The merchants needed the natives for crews, entertainment, and sandalwood harvest, while the missionaries urged the Hawaiians to stay home, plant crops, and save their souls.

Before sailing away, the *Wellington* crew emptied the dregs of the ship's water barrels into Maui's streams. Mosquito larvae squirmed in the water, surely visible to the naked eye. Some accounts say that the crewmen, upset that the missionaries had convinced the Hawaiian women not to sleep with them, dumped the insects out of spite. Or maybe the Hawaiian government was too deep in debt in the sandalwood trade and the frustrated merchants decided that if they couldn't recoup their money, at least they could share the misery.

The revenge motive seems farfetched and reeks of myth. Innocence or indifference is a much more likely reason for the mosquito's introduction. But a small voice says, "Why not?" Sailors, furious when missionaries discouraged native women from visiting the ships, attacked the house of the Reverend William Richards in Lahaina repeatedly in 1825 and 1826. Crewmen assaulted the governor and stormed the jail in order to release women held there. In 1827 the *John Palmer* blasted the mission with a cannon.

In comparison to this violence, the scattering of mosquitoes might have seemed like the throbbing of unsatisfied lust—sleep-wrecking, unbearably irritating, but ultimately harmless.

⅄ ⅄

THE mosquito larvae wriggled in these new pools on Maui, not as ideal a cradle as barrels of stagnant water, but not lethal either. Warm air buffered their arrival, and the water crawled with food. Segmented like an earthworm, each larvae displayed an enlarged thorax and head that disrupted the sinuous shape. Sweeping in bacteria and other microscopic organisms with their brushy mouthparts, they grew and molted, pushing through larger and larger larval forms.

Eventually the infant bloodsuckers turned into pupae, curved like bean sprouts. Here they didn't eat, just breathed and swelled. Inside the newly hard shells they developed wings, legs, and mature sex organs, then finally broke through the casings, splitting them down the back, stepping out as adults. A few moments after emerging, one rested on the surface of the water or on a

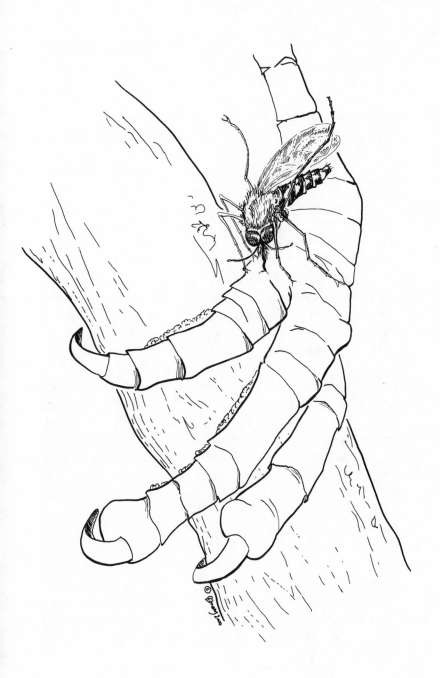

scrap of pupal case, slim and almost graceful, six long legs supporting the frame, drying in the warm air.

In the first days, trying out her new wings, she fed on nectar and fruit juice, like her male counterparts. The sugar energy readied her for mating almost as soon as she left the water. But after copulation she needed to eat in a way that would push her eggs to develop and mature; she looked for blood.

The mouth of the female mosquito is deceptively simple. The basic needle encases a toolbox of appliances: barbs to pierce the skin and capillary, a syringe to inject saliva that dilutes the blood, and a feeding tube to suck up the meal. Humming through the night forest, she sensed blood, taking in the smell, the carbon dioxide wafting from exposed skin. Near a village she might have found a colony of rock doves, introduced in the 1790s, dozing on the branches. Probing one dove until she discovered a vulnerable spot, the mosquito then shot her lancets through the skin and into a capillary and eased the feeding tube and syringe in alongside. As the blood pooled and the saliva thinned it, she fed until her insides bloated or the dove shivered to dislodge her.

With the blood-fed eggs maturing, the female searched for a spot to set them. Any basin of still water would do. After sending out a raft of several hundred eggs, she zipped back through the trees, looking for blood to nourish another batch. This time the mosquito might have sniffed out an amakihi, a greenish yellow bird native to Hawaii, sought skin on the foot or the corner of the beak as the bird shifted in sleep, and pushed the needle mouth in.

✳   ✳

HELPED by wind and more and more ships, mosquitoes spread to other islands. Twenty years after the *Wellington* had emptied its barrels, an island-hopping traveler named Chester S. Lyman stopped at Kealakekua Bay on the island of Hawaii and stayed overnight with a missionary. His joy in the morning gives an indication of what the rest of the trip was like. He wrote, "A kind greeting, a shower bath, a change of raiment, an excellent supper and after family worship a good bed free from fleas and mosquitoes were all very desirable and delightful." Another visitor, in 1868, commented on the influx of exotics in general: "The bad and the good have been introduced into that country in very equal proportions—the most beautiful trees, loveliest flowers, and delicious fruits, with the most annoying insects and loathsome diseases. . . . As for mosquitoes and white ants, words could not do them justice—the former devour you and the latter devour your furniture and books."

By the turn of the century, mosquitoes had evolved into stock scenery for dramas about Hawaii, like tropical breezes and coconut groves. Characters chided one another for discussing the insects; it was as boring as talking about the weather. The pesky bugs provided a dash of authenticity to Hawaii-based travel stories with titles like "Halcyonian Hawaii" and "Love-Life in a Lanai." No one remembered where they had come from.

꙼   ꙼

THE Hawaiian Islands, crests of submerged mountains far taller than Everest, created by lava piling up from the ocean floor, were first colonized by storms. Wind and waves brought their offerings: a pinch of fern spores, a wave of beetles, a flock of geese dazed by a hurricane. Estimates say a new species may have arrived only once every hundred thousand years, but there was no hurry.

About three million years ago a handful of small, adaptable birds arrived in the Hawaiian Islands. At least twenty-five hundred miles from the rest of its kind, the species mutated and shifted, changing so drastically that it is difficult to tell what it originally was, where it came from, what ancestor on the mainland can claim it. Time, like a sculptor with only one shade of

marble, set to populating the island with birds in all sorts of fanciful shapes. In a process called adaptive radiation, each evolved to fit a certain niche, becoming the best at gathering a specific kind of food or living in a certain habitat. The blunt, curved beak of the Maui parrotbill could easily snap the branches of the koa trees and drag out insect larvae curled inside. The blush-colored greater koa finch razored open the pods of the same tree with its sharp bill. The slender arched beak of the iiwi allowed it to perch upside down and sip nectar deep inside long-necked flowers. One of the most imposing beaks of all belonged to the grosbeak finch, a clamp backed by powerful muscles that cracked tough dried naio fruit to get at the seeds. All twenty-three species were honeycreepers, descendants of the first windblown colonists.

One of these variations is the amakihi, with green-tinged yellow feathers and a medium-size beak, small enough to fit in a tin cup. More of a generalist than relatives that wed themselves to one plant, the amakihi uses its modest-size beak to probe tree trunks for insects, search leaf crevices for spider eggs, and pierce the necks of flowers to get at their honey. Comfortable in a variety of climates, it lives in wet forests and dry ones, in areas that stretch from 250 feet to 10,000 feet above sea level. Six varieties live on six different islands, each altered slightly to fit its home's uniqueness.

Chance and isolation left Hawaii with no endemic land mammals save a bat, no endemic reptiles or amphibians; the diversity of native plants and birds was its showpiece. But with all the concern over resources to be exploited and souls to be saved, Hawaii's native species didn't get much attention, though they were startlingly unique. Moreover, the few records taken down by early naturalists were lost in shipwrecks, destroyed by fire, or simply misplaced. It wasn't until the end of the nineteenth century that the first comprehensive accounts of Hawaii's native plants and birds were available to the casual reader. A sense of this discovered abundance can be glimpsed through the eyes of Dr. R. C. L.

Perkins, a scientist who noted in 1892, "When I first arrived in Kona, the great ohia trees at an elevation of 2,500 feet were a mass of bloom and each of them was literally alive with hordes of crimson apapane and scarlet iiwi, while, continually crossing from the top of one great tree to another, the oo could be seen on the wing sometimes six or eight at a time."

This richness proved localized and short-lived. While naturalists slowly picked up their pencils, some birds in the island forests panted and lost their balance. Movements slowed to listlessness, and internal organs swelled. Other birds hopped along the ground, hobbled by sores twisting their toes and wing joints. A few that Perkins shot as specimens had diseased feet and faces, lesions that in some cases crippled them severely, signs of bird pox. Observers noted that Hawaii seemed to have more dead birds than did other places. Forests stood silent. Hawaii's birds were disappearing faster than they could be named.

The mystery didn't stump anyone at first. Reasons the birds might be dying abounded. Far before Captain Cook set foot on the beach, the Polynesians brought rats, dogs, and pigs with them when they arrived and settled the Hawaiian Islands between A.D. 400 and 600. Large predators in an environment innocent of them, the hungry mammals made quick work of species of flightless geese, ibis, and rails that disappeared before Europeans arrived, leaving only bones to be unearthed and labeled. Cook and those who followed added cows, goats, and sheep, which grazed patches of the island to bare earth. In the early 1880s mongooses, brought over to eat the rats, developed a preference for fledgling Hawaiian crows instead. Rabbits, brought to Laysan Island by a man who wanted to start a canning business, picked the island clean, leaving little habitat for the Laysan Island rails, millerbirds, and honeycreepers. House cats gobbled down whatever birds they could catch.

As more settlers arrived, villages overflowed with chickens and domestic geese. Starlings, house sparrows, and mynahs perched in Hawaii's trees, as did Chinese thrush, Japanese white-eye, and the Guam edible nest swiftlet, just a few of the more than one hundred nonnative birds living on the islands. These jockeyed with the Hawaiian birds for food and nesting sites, thriving in stands of imported plants that the Hawaiian birds couldn't eat. Florida blackberry grew up in thorny patches, banana poka vines wound themselves around forest trees, and lofty miconia plants shaded out the understory, replacing plants the native birds depended on. Each new boatload of visitors brought more seeds, pets, and stowaways.

But none of these could completely explain the eerie quiet in groves that should have been ringing with sociable chatter, nestling begging cries, and blustery squawks of territorial defense. In 1902 H. W. Henshaw walked in silence through lowland forests, even while bushes of native plants were in full bloom. With all these nectar-filled flowers, he couldn't credit the idea that the birds were dying off for lack of food. He wrote, "It is more reasonable to conclude that the former inhabitants of such tracts have abandoned them for the more profound solitudes higher up than that they have perished from such slight causes. However, even the abandonment of forest tracts under such circumstances seems inexplicable, and the writer can recall no similar phenomenon among American birds."

Henshaw put his finger on a pattern that became only more pronounced as the twentieth century unfurled. The nineteenth century's declining species plunged to extinction in the twentieth. Birds slipped through scientists' fingers like water, some vanishing even as they were first described, making the job of Hawaiian ornithologist a painful and depressing one. The koa finch and its razor bill disappeared; so did the naeo-splitting kona grosbeak, the black mamo with its prized yellow tufts, and the ula-ai-hawane that lived off the native palms. The astonishing diversity of the honeycreeper family, millions of years in the making, was halved in less than two centuries.

While lowland forest emptied out, remnant clusters of native birds retreated to higher and higher elevations. Graphs of bird populations showed them huddled on islands on top of the islands, as if the Hawaiian chain were slowly being submerged by waves only visible to the oo, the iiwi, and the apapane. Death, it seemed, was a mountain climber, inching up higher every year.

One clue to the mystery of the disappearing birds was uncovered by Ronald Ross, a doctor with the Indian Medical Service. Sweating away in Calcutta in the 1890s, he dissected mosquito after mosquito, searching for the cause of human malaria, before finally pulling out the salivary gland, packed with malarial parasites waiting for the next leg of their journey. Using sparrows and pigeons and larks, he traced the life cycle of avian malaria, and by association, human malaria, and proved the disease is spread by mosquito.

Through microscopes, he and colleagues pieced together a previously invisible side of the mosquito story, one reaching far beyond itching and annoyance. While biting a bird infected with avian malaria, a mosquito drinks in thousands of male and female malarial parasites encased in red blood cells. In the mosquito's stomach the parasites embark on the next stage of their life cycle. The males break out of the blood cells, develop spindly arms, and swim toward the females. They join, creating fertilized eggs which slip under a membrane on the surface of the mosquito's stomach to ripen and begin churning out thousands of long malarial sporozoites. These migrate into the salivary glands, waiting for the mosquito to bite.

When the mosquito feeds again, sporozoites surge into the second victim's capillaries. They first head for the cells of the bird's internal organs, then navigate into the bloodstream, attacking red corpuscles and moving inside. They grow and multiply, feeding on hemoglobin, repeatedly bursting out and colonizing new blood cells, until they change into male and female. Here, still incased in the corpuscles rocketing through the bloodstream, they are ready to be whisked away to their next destination by a hungry mosquito.

Domestic fowl or migrating ducks may have brought avian malaria to Hawaii before 1826, but the parasites had no place to go. Without a way to get from one bird to another and an insect to incubate the fertilized eggs, the malaria couldn't reproduce. But with the arrival of the mosquito, and flocks of chickens scratching outside their coops, malaria could complete its life cycle, and the piece necessary for the spread of the deadly disease snapped into place.

Like King Kamehameha II and Queen Kamamalu, and the Hawaiian natives who succumbed to smallpox, leprosy, and a tide of other European ills that washed over the islands, Hawaii's native birds had no defense against malaria, a disease that often brushed by other birds lightly. Honeycreepers died quickly, consumed by the invaders. While the release of malarial parasites flushed some birds with fever, others suffered from bird pox virus, also carried by mosquito. The whalers may have thought they were stirring up big trouble by releasing a tiny exotic insect, but the real damage came from exotic organisms too small to see.

But why did the remaining birds cling to the heights? In the 1960s scientist Richard E. Warner thought he had the answer. He noticed that the scraps of native bird life seemed to exist just above the line marking the top of the mosquitoes' range. All the birds living at the lower elevations, he reasoned, came into contact with mosquitoes and died of avian malaria and bird pox, among other diseases. The boundary between safety and infection, which he first placed at about 1,970-foot elevation, varied from island to island, depending on the habitat, but inched higher as settlers arrived. As people built houses and opened clearings in the forest, female mosquitoes found hospitable sites at higher and higher elevations and set their eggs in cattle troughs, pig wallows, ditches alongside roads, and tanks that collected rain for drinking water.

To test his theory, Warner captured Laysan finches from Laysan Island, one of the few places that had remained mosquito-free. He transported them to Honolulu in cages wrapped in cheesecloth to keep out the mosquitoes. For two months the finches stayed healthy, singing, and eating the raw eggs that made up a large part of their natural diet. Two weeks after the scientists removed the protective swaddling and opened the windows to let insects in, crippling lesions erupted on the birds' feet, mouths, and wings—signs of bird pox. Further experiments showed that Laysan finches and birds living at higher elevations quickly succumbed to avian malaria as well when exposed to mosquitoes.

Other scientists found that nonnative birds infected with malaria had many fewer of the parasites in their blood compared with native birds, consistent with the idea that birds from countries with histories of malaria would have developed immunity. The largest numbers of infections occurred in the zone where the low-elevation mosquitoes and the high-elevation birds overlapped, where unexposed birds and infected insects came into contact.

Over the years, as more ships and eventually airplanes stopped at the Hawaiian Islands, two day-flying mosquitoes—*Aedes aegypti* and *Aedes albopictus*—and another night-flying mosquito—*Aedes vexans nocturnus*—provided round-the-clock annoyance. The Hawaiian government launched control efforts, mixtures of insecticides and biological control, beat back some of the swarms, and gave the human and bird populations some relief. Still, a woman who grew up in Honolulu describes her mother lighting a match at the Buddhist altar in their home, holding it to the incense, and dousing it in a jar of water kept by the shrine. In the jar mosquito larvae squirmed.

✻   ✻

DOCUMENTING this relentless subtraction and recording the birds' retreat to forty-five hundred feet were undeniably bleak work. But in the bird watchers' and scientists' notes, another pattern, more heartening, began to emerge.

Of the amakihi taken from the slopes of Mauna Loa volcano on the island of Hawaii—the adaptable bird that pierced the base of flowers to drink the nectar and plucked spiders out of crevices of bark—all but one of those experimentally infected with avian malaria survived. Also, a woman spotted a flock of amakihi visiting blossoming vines near her Honolulu home, well down into the mosquito danger zone. More recently a biologist with the University of Hawaii tested amakihi in lower elevations on Oahu and found them all, astonishingly, malaria free.

Compared with some other species that migrate between elevations, flitting between mosquito-infested and mosquito-free zones, the amakihi are relatively sedentary. Living in an infected area, they would be constantly exposed, a factor that might help them develop immunity. Since amakihi have short generations—one-year-olds are able to breed—they may have gone through enough genes so that a mutation resistant to malaria emerged, letting the bearers live and lay eggs that contain the same survival secrets. The new results set scientists scurrying to isolate the gene that gives these amakihi immunity and use it to create a disease-free strain. Even without the biologists' assistance, forces of natural selection may be now preserving birds that are resistant to malaria and bird pox, thwarting even anger made manifest in the form of a sprinkling of insects on a welcoming shore.

This clinical gamut of blood tests and DNA analysis has led those studying Hawaiian birds to a sensation they don't often get to experience, hope, defined by Emily Dickinson as "the thing with feathers—That perches in the Soul." An apt description perhaps of the emotion rising in an ornithologist watching amakihi bicker over lantana flowers, despite the low elevation, the gathering dusk, and the mosquito's bloodthirsty hum.

CHAPTER FIVE

AN ARTIFICIAL WEDDING

THE trip back up the Erie Canal was quiet in comparison with the dizzy energy of the days before. The downstream float of the *Seneca Chief* marked the opening of the completed canal on October 25, 1825, and the state was wound up with pride. New York Governor De Witt Clinton rode on the barge, making speeches, gathering praise. Flower garlands spanned the canal. Cannons roared welcome. Barges, including the *Noah's Ark*, which carried representatives of western animals from insects to bears, joined the procession from Buffalo to the ocean. Now, two weeks later, fading echoes of cheers, gun salutes, and the pop and sizzle of fireworks made the mornings seem only more hushed. Sore toes and arches reminded some of a long night at the Grand Canal Ball. Others savored the remembered smell of Canal Beef, cuts of fat inland steer, carried to the Atlantic coast by barge. One or two recollected the sight of City Hall glowing with a thousand wax candles or a glimpse of the maple sugar canalboat that floated on lake water, the centerpiece at one of the parties by the shore. The climax of the cruise downstream had been "The Wedding of the Waters," when Clinton poured two barrels of water from Lake Erie into the Atlantic. To amplify the grandness of the gesture, a Dr. Mitchell had collected samples from the Elbe, the Ganges, the Nile, the Amazon, the Columbia, the Thames, and the Seine, among others, to sprinkle into the ocean as well. The barrel of sea water they took back to mingle with Lake Erie was, like the voyage up canal itself, an afterthought.

But they dutifully dumped the Atlantic fluid into Lake Erie from the cask marked "Neptune's Return to Pan" in gold lettering. As the drops merged into the body of the lake, releasing microscopic creatures and salt, the weary passengers witnessed a symbol with even more resonance than the explosion of the thirty-two-pound cannonball that had launched them on their journey.

A few miles away on the other side of the Niagara River, the Canadians absorbed the reverberations. They were building a canal too and knew the importance of their neighbors' having a completed connection to the sea. At the end of 1825, as the party for the opening of the Erie Canal sputtered to an end and the business of hauling goods from inland to ocean began in earnest, the Canadians had plans and surveys and opinions of engineers. They had a government-approved private company interested in building the canal, the Welland Canal Company, and they had an opening ceremony as the first shovelful of dirt was removed from the spot where the canal would be. That was about all.

Both countries were seeking to dig their way around a problem, and the problem's heart was stubborn geography. Lake Erie and Lake Ontario nestle just next to each other, almost touching, like two slugs about to kiss. To the north and east, Lake Ontario drains into the Atlantic Ocean through the slow vein of the St. Lawrence River. To the south and west, Lake Erie touches the fan of the other Great Lakes: Huron, Michigan, and Superior. Water from Lake Erie races down to Lake Ontario through the Niagara River, which also serves to divide the United States from Canada: Buffalo, New York, on one side, Fort Erie, Ontario, on the other. This should be the final link in a liquid chain connecting the Midwest and the sea, but there is one small obstacle.

Niagara Falls's ability to inspire awe and terror made it one of the first scenic spectacles in the United States. Honeymooning couples came to stare at the torrent leaping over the almost two-hundred-foot drop and heard in its roar an echo of their own passion. Thomas Moore saw God in the crash and flow. He wrote to his mother about the falls: "I felt as if approaching the very residence of the deity; the tears started in my eyes." Charles Dickens found peace with a capital *P* and wrote: "Niagara was at once stamped on my heart, an Image of Beauty; to remain there, changeless and indelible, until its pulse ceased to beat, forever." Other onlookers discovered in the tumbling droplets a lesson in the power of nature, evidence of geologic time, and, even in the early 1800s, a tourist trap. The falls were many things to many people. One of the things they were was in the way.

At stake were the grain and meat and metal of the inland portions of the United States and Canada. Both New York City and Montreal envisioned themselves as the dominant trading center in the East, and a canal would only boost their claims. Before the railroads crisscrossed the country, water was the most important method of transport. With a waterway, goods would practically float themselves to market, while without one, the same trip took weeks of dragging over bad roads with tired animals, an expensive and backbreaking proposition. If businesspeople could take goods from Lake Erie to Lake Ontario, they could sell them all along the East Coast. But the two lakes were separated by a change in elevation of more than three hundred feet, much of which occurred in the steep drop of those pesky falls, shooting off spray and rainbows and romance. No boat could pass over them, and neither could any goods, so farmers and middlemen were left carting their loads between the lakes over miles of uneven terrain.

New York State's strategy to defeat the falls involved carving a trough from Lake Erie to Troy, from which point boats could ride the Hudson River down to the Atlantic. This route, though long and convoluted, kept goods in American territory. In contrast, all Canada had to do to establish a passageway in its territory was cut through the small Niagara peninsula and sidestep the falls. This plan was best summed up by Sebastian Nauban, a traveler who had similar notion in 1699. He looked at the torrent of water and declared, "Niagara Falls . . . is tremendously high, but there is nothing which cannot be corrected by man."

One of those watching the American's canal fervor most closely was a failed shopkeeper named William Hamilton Merritt. His mill on Twelve-Mile Creek, a slender body of water that fed into Lake Ontario, didn't always receive enough water to run, and he needed it to quite badly. In the prosperity after the War of 1812 he'd floated several business ventures, and within a few years the prosperity had drained away and left him bankrupt. If he could somehow connect Twelve-Mile Creek to the Welland River, the water would flow by, and his mill would run as much as he needed. Since the Welland River connected to the Niagara River above the waterfalls, his plan would incidentally provide an important national service. Merritt, noted for his energy, enthusiasm, and missing sense of humor, was eager to get his project under way. He had no qualms about linking his personal needs to patriotism. In speeches about the Great Lakes and the access a canal would offer, he was soon declaring, "These seas affording the most beautiful and commodious means of internal communication ever seen. . . . It is truly a national object."

With the help of influential friends, Merritt put together the Welland Canal Company and got permission from the government to gather funds. He surveyed a route and hired work crews. He borrowed money, and his men started to dig. The walls of the deepest cut in the canal collapsed, as digging revealed their sand foundations, and his engineers revised their recommendations. They ran out of money and went looking for more. They built dams from branches and locks from wood. (The lock is the solution to the problem posed by a waterfall. As the water between the lock's two gates is raised or lowered, boats move up or down as if borne by a cableless elevator. They gently rise and sink with mechanized grace, rather than the plunge and tumble of unfettered water dragged by gravity.) Year by year and mile by mile, the project snaked forward. Ultimately Merritt seems to have been caught in his own trap. Hoping to get the country to support his personal interests, he ended up sinking his personal interest in service of his country.

By 1829 the canal's first incarnation had been finished: a sprawling Y with its base on Port Dalhousie on Lake Ontario, one arm reaching into the Welland River, and from there to the upper Niagara, and the other extending to the Grand River for a short jaunt before touching Lake Erie. With forty locks, it was twenty-seven miles long, eight feet deep, and far over budget. On November 27 Merritt traveled the newly opened canal from its mouth on Lake Ontario to Buffalo. The boats were swathed in bunting, but they made halting progress through ice and logs. In spots they ran aground. It was no Erie Canal celebration—no flower garlands or maple sugar model boats embellished the scene—but ships could pass and water flowed through Merritt's mill. In referring to the trip, Merritt recalled the Americans' festivities and their barrels of water, noting that with the finished Welland Canal, "The artificial wedding of the Great Lakes of the west and north with the waters of the Ontario, and eventually with the St. Lawrence and the ocean, was complete."

Despite the headaches of planning and funding, it must have seemed like a miracle, all that water could do. Thousands of pounds of grain and ships broad and awkward slipped right through. Witnesses must have felt lighter watching the push of the current, as if just that moment they'd put down the goods that had to be carried from place to place. Now that the ditch was dug, they just needed to stand back and let the canal flow. This wasn't the sparkling romantic water of Niagara Falls; it was the hardworking grease of industry. The water rushed on, carrying all that floated over it and all that lived in it.

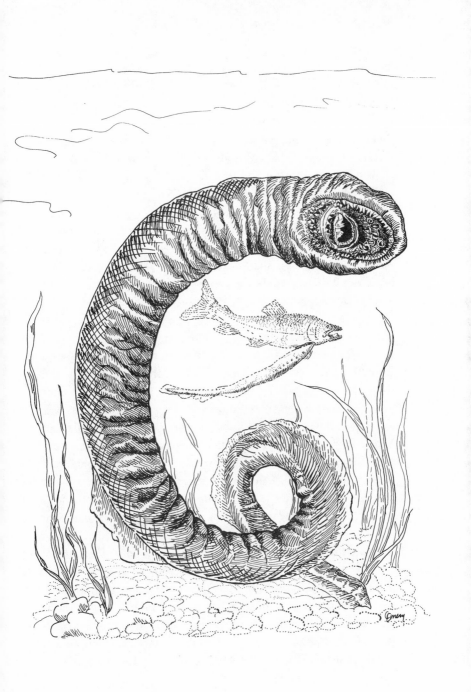

⚜   ⚜

THE sea lamprey is hard to like. Rings of needle-sharp teeth form the O of a mouth. The rest is all dark undulations, the body type that slips snakes and worms into so many nightmares, a stream-lined feeding tube. When the lamprey scents or sees its prey, it fastens its mouth to the skin and then slides over the body until it finds the spot it wants to enter. With a raspy tongue, it begins to scrape away the flesh, creating a vacuum in its throat to suck down the blood, tissue, and eventually bones and internal organs. Secretions from glands in the lamprey's mouth prevent the fish's blood from coagulating and keep the food supply liquid. Some have speculated that since the prey of the lamprey often dies, it is less a parasite than a predator. Just a very slow one.

The life stages of the sea lamprey resemble those of a more charismatic creature, the salmon. While many species of lamprey travel from riverbeds to the sea, the sea lamprey that frequents Lake Ontario is landlocked, so it migrates from river to lake and back again. In late spring and early summer, when the water temperature is right, mature lampreys begin to fight their way upstream. Traveling at night, they make one or two miles a day. When they find a good spot, chunky gravel with a bit of sand, water not too deep or too shallow, they dig out nests. Sand and rocks form a downstream heap, reversing the flow of the current over the eggs, forming a kind environment for hatching. As they tire fighting upstream, lampreys suck on to rocks or fallen branches with their mouths until they are ready to travel again. This mode of navigation prompted their Latin name, *Petromyzon marinus*, the first part of which means "stone sucker."

Ancient creatures, lampreys developed from jawless fishes that swam through prehistoric waters five hundred million years ago, as evolution strung the first few vertebrae to make a backbone. Two hundred and thirty million years ago, as conifers marched over the landscape, the ancestors of the Rockies were beginning to rise, and cicadas filled summer nights with their first buzz, a jawless fish died in what later became a coal mine, south of what later was Chicago. It was a ringer for the lamprey swimming in the Great Lakes today, though it had fewer teeth. The unadorned design persisted. No jaws. No bones. No scales. Simple and efficient.

As well as being a barrier to industry, Niagara Falls had been a barricade against aquatic species. The most zealous spawner could not crest Horseshoe Falls or American Falls with a 170-foot leap. As a result, species that lived in Lake Ontario had not yet made it to Lake Erie and the chain of other Great Lakes connected to it. The sea lamprey was one of these; it has a sprawling range spanning the Atlantic coast but was a foreigner to inland waters. It's unclear if this lamprey is native to Lake Ontario, but it was present there by the early 1800s. With the Welland Canal complete, however, all sorts of creatures discovered easy passage.

⚹   ⚹

THE Canadians were not content with their over-budget, marginally serviceable ditch. The canal was rerouted, deepened, and improved until it barely resembled its former self, always letting through more water, making the connection stronger, clearer. Interest grew. Control changed hands. It was as if Canada were an architect laboring over drawings of its future, worrying the original sketch, erasing and amending over the course of a century. From a project that began with shovels and wheelbarrows, the Welland Canal and its offspring sprouted a forest of machinery.

In 1845 the second canal retraced the route of the original but deepened the passage. In 1887 the third Welland Canal was completed, guiding water through twenty-six locks in a channel no shallower than fourteen feet. The fourth canal, opened in 1932–33, abandoned the old launch point at Port Dalhousie on Lake Ontario (chosen to shunt water past Twelve-Mile Creek and Merritt's mill) in favor of Port Weller, finally straightening the meandering curve of earlier attempts. In the new design water flowed to a depth of thirty feet through eight massive locks.

Who knows when the first lamprey set out to spawn, found its natal gravel pits disturbed, and kept going upstream, wriggling through new waters, until it became the first of its species in the fish-rich pool of Lake Erie? But in 1921 a fisherman in Merlin, Ontario, pulled a sea lamprey from his nets. Along with other strange fish he'd discovered in Lake Erie—a silvery lamprey, a rosy-faced minnow, and a miller's-thumb—he sent it on to the Department of Biology at the University of Toronto, which included the lamprey, with no comment, in its 1922 list of the fish of Lake Erie. Erie's shallow waters didn't appeal to the lamprey, but it was easy enough to slip from there into the deeper lakes of Huron, Michigan, and Superior. Soon anglers were unhooking whitefish with circular wounds on their sides and hauling in lake trout with two or three lampreys dangling, still sucking.

The speed at which the lampreys did their work is breathtaking. By the mid-1940s, only twenty years after they first slithered in, no one could ignore the effects on the Great Lakes. The parasites scraped away at lake trout, rainbow trout, and whitefish, their fervor pushing two varieties of chub close to extinction. In Lake Michigan the trout catch plummeted from more than 5,000,000 pounds in 1945 to only 500,000 pounds in 1949. Fishermen on Lake Huron, used to pulling in 1,720,000 pounds of trout, watched their haul dwindle to about 4,000 pounds over the course of twelve years. Losses topped five million dollars annually. In some years 90 percent of the lake trout in Lake Superior had lamprey scars. Not all the damage was immediately apparent. In one case, after a storm had stirred up the waters, acres of dead, lamprey-scarred trout rose from the bottom to the surface.

Roiled by the newcomer, the entire ecological landscape settled into a new shape. Lampreys gnawed at burbot, which in turn fed on sculpins. With their predators out of the way, the sculpins multiplied and ate more lake trout eggs, pushing them closer to the brink. Filling vacancies created by lamprey attacks, inedible alewives moved in, outcompeting more flavorful species and increasing the fishermen's disgust.

Ever looking for a market solution, the Fish and Wildlife Service proposed lampreys as a good source of vitamin D. Some cultures consider them a delicacy, officials urged. Emperor Vitellius of ancient Rome supposedly favored them served with pheasant brains, and King John of England died from eating too many, newspapers noted. But not all the encouragement in the world could make midwesterners crave lamprey stew. They slither. They're ugly. Infections and fungi riddle the flesh. In a show of public spirit, a daring biology teacher breaded one, fried it up, and had a taste. He said it wasn't bad but declined a second bite.

Scientists tried other tactics. Electric shocks, physical barriers to spawning grounds, lampricides, and releases of sterile males all helped fish populations rebound slightly, but the lamprey found ways to persist. About six hundred thousand still live in the Great Lakes, wriggling through the rocks of tributaries and heading into Michigan, Superior, Erie, Ontario, and Huron when the predatory urge hits.

৵ ৵

THE Welland Canal is now part of the St. Lawrence Seaway, a string of lakes, rivers, canals, and locks that allows ships to travel from Lake Superior to the ocean. These oceangoing vessels, lumbering more than seven hundred feet in length, make the original barges bobbing down the canals look like little maple sugar toys. The modern waterway towers over the Erie Canal, which never achieved a depth of more than twelve feet, and now attracts more pleasure boaters and historians than captains of industry. Since its opening in 1959 the seaway has allowed coal, wheat, sunflower seeds, salt, iron, and more than forty exotic species to pass through.

Some, like the lamprey, navigated their own course from ocean to river to lake. Others hitched a ride. When empty, big freighters weigh themselves down with ballast water to steady them against ocean swells. They inhale the fluid at one port and exhale it at another, an exchange of millions of gallons that makes a mockery of Clinton's little cask and Dr. Mitchell's vials. Crabs, mussels, and microscopic invertebrates find themselves whisked from China to England, from the Black Sea to the San Francisco Bay, inadvertent stowaways.

If the fisherman who pulled the first sea lamprey out of Lake Erie went out on the Great Lakes today, he would haul in all sorts of unrecognizable creatures. The ruffe protects its squat orange body with spines; the prickly dorsal fin, arching up like a crown, marks it as queen of the bottom feeders. The rusty crayfish, native to the southern waters of Tennessee, chases after mayflies and stone flies with its outsize claws. The round goby peers at the world from eyes on the top of its head, giving it the appearance of an overgrown tadpole. Feathery strands of Eurasian milfoil or the curling white blossoms of flowering rush would tangle in his nets.

Inspecting the hull of his boat, he might discover the zebra mussel, a mollusk native to the Black Sea, that may have been released into Lake St. Clair during the 1980s in ballast water from a ship traveling through the St. Lawrence Seaway. Though more benign in appearance, no bigger than a pistachio nut and modestly tucked between two striped shells, the zebra mussel is more despised than is the lamprey. Coating oars and anchors, massing on crustaceans and clams, shutting down waterworks, the zebra mussel intimidates through sheer numbers. Shipwrecks, pilings, entire lake basins are covered in a prickly blanket. Native mollusk populations are in decline, and diving ducks that eat the exotic mussels ingest contaminants along with their meal. Traveling through another canal, the mussel found itself in the Mississippi and extended its reach all the way down to Louisiana. The spiny water flea, white perch, and purple loosestrife have also taken up residence in the Great Lakes, changing the ecosystem forever.

✳   ✳

THE Northwest Passage, that golden link between Atlantic and Pacific, the ultimate canal from Europe to China, eluded explorers dreaming of Far East treasure. When it no longer seemed likely that the pounding surf or endless channel would appear just up one more drainage, they set out to carve a way through the land, if not all the way to the opposite shore, at least as far as possible. The goals changed from silk and spices to beef and wheat. They stitched lake to lake and wed salt water to fresh.

This passage, so long imagined onto maps, seemed the natural state of the land, confounded by mountains and waterfalls. On the eve of the Erie Canal celebration, as Clinton prepared his victory speeches, Samuel T. Wilson wrote a poem commemorating the event and its engineers, who helped not just industry grow but the water break free and realize its destiny. The verse appeared in the *New York Statesman* in 1825, and the final stanza crowed:

> Against immortal foes they stood
> Champions of freedom and of mind;
> And warring for the mountain flood
> In nature's fastness confined,
> Burst the huge bars, and rocky chain
> 'Gainst which for many a century,
> Their prison'd strength had dashed in vain,
> And set the world of waters free;
> Hills sundered, stubborn rocks were crushed,
> Mountains sunk down, and forests bowed,
> As to the parent sea they rushed,
> Uttering their voice of joy, aloud.

Ironically, 175 years later, scientists would be promoting these barriers, emblems of "nature's fastness," as the creators of biodiversity. Species are formed through separation and isolation as they evolve to fit the habitats where they are confined, building up defenses against their unique dangers, relaxing into their patches of safety. The barrier of the Pacific Ocean created Hawaiian geese from Canada geese. Glaciers along the Appalachian Mountains split mallards from black ducks. Utah landforms bred singular species of fish in the few pockets of water. Because of Niagara Falls, Great Lakes lake trout had no need to discover defenses against the sea lamprey and native mussels didn't have to figure out how to compete with zebra mussels. Many organisms staked their livelihoods on buckles and folds. The promise, like the peninsula, has been breached.

Water, though it will flow into any vessel and adopt any shape, has stubborn properties of its own. As waves of exotics pound against shores as far inland as Lake Superior and ruffe and goby settle into the muddy bottoms of Lake Huron and Lake Michigan, it's as if the willful ocean were moving inland to reclaim those first molecules of brine that took the trip upstream in a decorated barrel, poured into Lake Erie by Clinton and his crew.

11

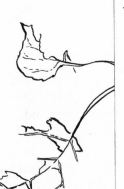

VICTORIAN

EMBELLISHMENTS

CHAPTER SIX

FOLLOWING SILK THREADS

EVENTS that summer unfolded with the grim predictability of a horror story. Though predating Hollywood B movies such as *The Deadly Mantis, Arachnophobia,* or *The Fly,* this tale featured all the elements of a hack screenwriter's fantasy. The mad scientist. A repulsive creature invading a suburban neighborhood. Women shrieking when they opened their closets. One slip, one tipped vial, and cities along the East Coast lived under swarming hoards of insect larvae that stripped apple and oak trees, wormed their way into houses, then turned and rose in a mad flutter of wings. Despite pesticides and predators, they spread, outbreak by outbreak, leaving a trail of bare branches and disgust. Eventually the story line veered beyond anything a cinema screen could hold.

In a modest house at 27 Myrtle Street in Medford, Massachusetts, in 1868, Leopold Trouvelot, an astronomer with a knack for natural history, was trying to breed a better silkworm. He imported several European gypsy moth eggs, planning to cross them with a North American species. If his efforts succeeded, he could jump-start the silk industry in the United States and boost his own fortunes as well. When several of the moths escaped one warm day in 1868, the plot ground into motion.

ஃ   ஃ

HOPING that stories stick to walls like paint or sink into the ground like motor oil, I go looking for 27 Myrtle Street, Trouvelot's old home. With me I have a street map from 1895, a photograph of the house, and a thick book called *The Gypsy Moth*. Written in 1896 by two zealous scientists of the Massachusetts Board of Agriculture, the five-hundred-page tome describes political, historical, and biological aspects of the invasion. Charts detail the number of gypsy moth caterpillars killed each year by government officials. Diagrams display every available method for spraying insecticides. Page after page shows photographs of bare shrubs and leafless trees. The book is a hymn to the belief that knowledge is ammunition, a testament to obsession.

After winding through the streets of Medford, following the command of one-way signs, I light on Myrtle. The quiet neighborhood near the Mystic River teeters on the edge of disrepair. A boy roller-blades down the sidewalk in too-big overalls, his baseball cap on backward. He zooms by a man in a wheelchair going the other way. Dogs behind chain-link fences growl threats at passersby. Weeds push through cracks in the asphalt. A wind chime tinkles over distant traffic.

The houses on the left side tick off in numerical order: 35, 31, 29, 25. There is no 27. Possibilities leap to mind. Trouvelot's angered neighbors burned it down. The house, infested with moths from cellar to ceiling, was dismantled by the state. Or more recent owners simply changed the number. My photograph is no help. All the houses are identical, and the basic blueprint could have sprung from a child's pen: a boxy front, a peaked roof, a door to the left with several steps leading up to it, and four windows to the right, two on each floor. Within this outline, each has small adjustments, shifts, and ruffles in response to the owner's personality. One house is pale yellow; another, pale green. Some have brown wood showing through chipped white paint; others display renovated windows, pulled to the center, and the sun shines on new glass. One yard is tightly cropped and fenced. In another, a knee-high Virgin Mary, robed in blue, keeps watch over daisies and nasturtiums.

The house in the 1895 photograph has wooden cutouts, like waves, running along the front of the upside-down V of the roof. Surrounded by a picket fence, the yard contains an arbor with vines massed on top. In front of the house stand a hitching post and a thin tree, staked on both sides like a splinted arm. A tree in the yard looks even worse. Leaves fluff one or two branches, but the rest are bare sticks. The gypsy moths had feasted, then moved on. Now, decades later, the trees have recovered, or been replanted, but they are not so noticeable beside the telephone poles marching down the street, lacing the sky with wires, or the Fords, Mazdas, and Toyotas parked against the curb.

For my own satisfaction, after traveling all this way, I decide that Trouvelot's house could have been 29. Slightly self-conscious about staring, I lean against my car with the expression of someone attempting to locate the address of a good friend. With information gathered from side glances at number 29, I try to strip away this twentieth-century veneer of cars and telephone wires and see what might have gone on within those walls one afternoon in May 1868.

Somewhere, maybe moving into the kitchen, a man searches for his pen. He has a crooked nose, and his eyes are close-set, dark with intelligence. Black hair sweeps up off a pale forehead, and a beard buries the lower part of his face, but under all that hair sits a rather delicate mouth. He is talking to himself as he peers at a cocoon, tilting it into the light. The transformation of a caterpillar into a fragile pair of wings never ceases to fascinate him. Earlier he sliced the cocoon open and replaced one half with a small window so he could watch the moth take shape, and it does, right now, but he can't find a pen to record it.

The kitchen table is covered with his experiments: deformed moths, whose wings he pricked when they were newly hatched and fluid-filled; silk reservoirs he cut from caterpillars to see how far they would stretch; pupae in a bottle. He has just come from his other laboratory, the five acres of shrubs and weeds behind his house, where, under nets, he keeps thousands of moths and caterpillars captive. Still warm after chasing the hungry robins off the bushes, he feels sweat beading on his face. Perhaps he wipes his forehead with a silk handkerchief.

Fashionable ladies of the time are swathed in silk, from bonnet to shoe: black silks for mourning, silk velvets for opera mantles, intimate silks for the wedding trousseau, rose silks, pink silks, maize silks, silver, gray, and blue silks, light-striped silks for spring, cheap checked silks for daily wear, grand silks with the new satin stripe, bonnet silks, sewing silks, walking, evening, and dinner silks, silk for sleeveless jackets (new for May!) in Duchess and Walter Scott styles. Hundreds of yards of the lustrous fabric pile in warehouses and shops waiting for consumers to come, to buy, with appetites almost as voracious as those of the caterpillars themselves.

With a new tax on imported silks, and disease running through French and Italian silkworm populations, the time seems right for an American silk industry. Magazines like *Scientific American* urge readers to breed silkworms and feature frequent articles on the topic. Pasteur is intrigued; Ezra Stiles, the president of Yale College, spends forty years studying the moths. In California, where hopefuls raise millions of trees and cocoons, a man who envisions a silk community proposes carving up a ranch near Los Angeles and selling ten-acre lots to families that will plant mulberry trees and raise silkworms. The state fair displays California-grown silk fringe and demonstrations of silk reeling and cloth weaving. The new business is economical as well. In 1871 W. V. Andrews notes in *Scientific American* that "The labor of a few old men, or women, or even children, is sufficient for the purpose. The cost is therefore trifling."

In Massachusetts a textile industry based on cotton and wool flourishes, with the mills in Lawrence and Lowell churning out yard after yard. Power looms and new machines that weave lush

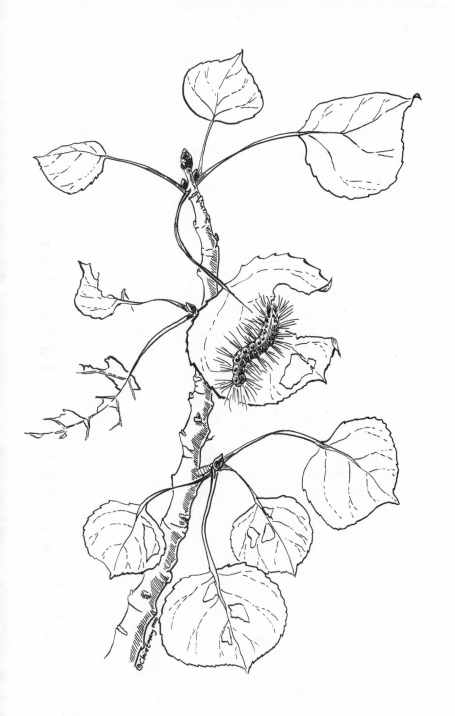

velvets and complex patterns revolutionize the industry, cracking open a market for silk. Mills and foundries churn and roar the country into a new shape. For manufacturers, this is a time when production is limited only by ingenuity, market smarts, and the willingness to risk. Young farm women and immigrants are only beginning to organize and object to being treated like machines.

But the American silkworms aren't cooperating, and imported raw silk from China still feeds the American industry. Most American species spin silk that proves dull and difficult to card or create dense cocoons that are tightly stuck together and weak-fibered. Raising silkworms in the Northeast is particularly chancy because the insects often die in the cold, and one *Scientific American* author argues that introducing new species is of "so much prospective importance, that I shall devote the remainder of this article to the consideration of whether *Yamamai* and *Pernyi* may not be naturalized here." He is not the only one dissatisfied with native species. The Council of Paris Acclimatization Society imports silkworms from North America, while Japanese silkworms munch on oaks in Austria.

Trouvelot's wide-ranging and restless mind seizes on this problem. Maybe he's drawn in by waving antennae and powder-dusted wings. Or maybe he imagines silk threads will carry him across the river to a nicer house in Cambridge, where the big scientists peer through big telescopes. Either way, he's experimented for years with *Telea polyphemus*, a North American species whose caterpillar spins a strong and glossy thread but is notoriously difficult to raise. Now he has new hopes. Amid the vials, cocoons, and moth parts lie the eggs of the European gypsy moth, whose current classification, *Bombyx dispar*, places it in the same genus as the silkworm. When crossed with the North American moths, the gypsy moths might produce a heartier breed, easy to raise, and able to withstand the New England winters. Mind alive with possibilities, Trouvelot takes a last look at the emerging moth and spots a pen wedged at the back of a shelf. And then . . .

Here my imagination yields to three different accounts of the accident. The eggs, resting on a table, blow out the window with a gust of wind, or the eggs, enclosed in a vial, are knocked from the windowsill and set free, or wind rips a hole in the netting covering Trouvelot's experimental plots. But somehow, the first small spheres drop to the grass.

The eggs nestle there for days or weeks, growing dark as the embryo swells. One day their first instinct pricks to life, and they start to chew, gnawing a hole in the egg casing, then enlarging the circle, until only scraps of the top and bottom of the shells remain. Once free, they rest for a day, or two, or three, nibbling at leaf hairs. Eventually the tiny caterpillars turn and move toward the light until they stumble across another leaf.

If it is a black currant bush, a turnip plant, a weeping willow, a red maple, a black oak, a white pine, or any other of more than four hundred plants found in Massachusetts that they find edible, they feed, starting a circle in the leaf and then chewing around it, bite by bite, until holes spread like acid on cotton. In Trouvelot's yard they find a rosebush or maybe an oak. When one tree is stripped, they drop down on a silk thread and let the wind launch them to another patch of green. When older, they will eat along the edge of a leaf, traveling around again and again, jaws working methodically, until all that's left is a spine.

As they eat, they strain their skins and molt, pulling their heads from the old casing, then pressing backward until the casing tears. Crawling out of this crevice, they resume the search for food. They molt, grow, and molt again. Hairs sprout from the red and blue bumps that file down their backs. Gray bodies end in black heads marked with fingers of white, skunklike. Traveling on many legs over twigs and limbs, they stop occasionally to rear back and test the air.

A few months after first hatching, they seek a sheltered spot, like the eaves of the astronomer's shed or the corner of his fence. In three hours they pupate, and over the course of ten to twelve more, they shift and change, growing new structures and losing the old. Finally they emerge, wet and exhausted. A male moth dries off nut brown wings and feathery antennae tuned to the female scent. A female, larger, with thin, curved antennae, flaps light wings painted with black spots and gray scallops. Weighed down with eggs, she is too heavy to fly, but the males are already cruising in a zigzag pattern, trying to catch her smell. They are gathering around her. She has only to wait.

After mating, the female rubs her abdomen against the tree trunk or fence post, leaving fine hairs in the bark, then deposits her eggs, layer upon layer, hour after hour, gumming the mass together with abdominal hairs and her own glue. Several days later she completes her egg mass, each yellow mound capable of hatching up to one thousand new caterpillars the next summer.

Back in the kitchen, Trouvelot must shiver when he hears the "clink" of broken glass or finds a tear in his backyard netting. Having witnessed the ravages of a native silkworm, he has noted, "What a destruction of leaves this single species of insect could make if only a one-hundredth part of the eggs laid came to maturity." Panic knifes through him as he searches the grass, and he tells everyone he knows to be wary, but his mind can't possibly encompass the destruction that will follow, the acres defoliated and millions spent on pesticides, any more than he can picture a man on the moon.

Ten years later Trouvelot was gone, and his warnings were forgotten. Mr. William Taylor moved to number 27. Exploring the shed in the back of his new property, he found the walls crawling with caterpillars. He quickly sold the enclosure, but the insects persisted. "In their season I used to gather them literally by the quart before going to work in the morning," he wrote.

The neighbors, first those who lived close to the astronomer, then those who lived farther down the block, began to notice something seriously wrong. They told story after story of discomfort verging on the unbearable, and the devoted authors of

The Gypsy Moth took it all down. Next door to Trouvelot's old home, Mrs. William Belcher couldn't ignore the newcomers: "They were all over the inside of the house, as well as the trees." Another neighbor noted, "The caterpillars would get into the house in spite of every precaution. We would even find them on the clothing hanging in the closets."

Several houses away, in 5 Myrtle, Mrs. D. W. Daly waged a personal battle against the moths and their larvae. "I spent much time in killing caterpillars," she reported. "I used to sweep them off the side of the house and get dustpanfuls of them. At night time we could hear the caterpillars eating in the trees and their excrement dropping to the ground."

Then, in 1889, things started to get really bad.

Having denuded Myrtle Street, the gypsy moths spread block by block and flourished with the new food supply. As the caterpillars fell from the trees, one man covered his head with his coat and ran to catch the train to work, but at the station he found his jacket lining squirming with them. A woman described the embarrassment of having to sweep off her front steps every time guests stopped by, so they wouldn't arrive with their shoes covered with insects. Masses of caterpillars blackened housefronts. Days were filled with raking and burning piles of leaves infested with caterpillars, pouring boiling water over caterpillars marching along fences, picking caterpillars off the wash and from underneath the pillows. Many people stayed inside on hot summer days to avoid insects in their hair and clothes. The stench of dead bugs hung over Medford like smoke, intensifying in the summer heat. The noise of the feeding insects sounded like "the clipping of scissors," "a breeze," or "the pattering of very fine rain drops," and whoever heard them knew she would wake up to a yardful of skeleton plants. Some tried to move out of the neighborhood but were invariably met by questions about the trees, leafless and bare in June. One glance at a caterpillar crawling across a trouser cuff, and the prospective house buyer was gone.

As the outbreak progressed, the state launched its own attack. In 1890 the Massachusetts legislature gathered a commission to halt the progress of the gypsy moth. Teams of men descended on egg clusters and burned them, sprayed them with acid, doused them with oil, and coated them with tar and varnish. Individual trees of sentimental or commercial interest were girdled with tar paper, in hopes that the caterpillars would stick, unable to cross. Others were wrapped with burlap, which lured caterpillars to its shelter so that managers could remove and destroy them. They sprayed Paris green, London purple, and other arsenic-based poisons over acres of hatched caterpillars, attempting to kill them before they bred. Explorers scoured Europe and Asia for an insect that might prey on the gypsy moth, and in 1909 researcher George Clinton returned from Japan with a fungus believed to cripple the larvae, but it vanished into the Massachusetts forests, and the moths continued their hungry progress.

Meanwhile, in his new home in Cambridge, Trouvelot shifted his gaze upward and pursued his true passion, astronomy. At his own observatory and at Harvard's, he watched fiery arms fling themselves from the sun, reaching for a distance of three hundred thousand million miles, flaring and disappearing in seconds. He spent nights gauging the transparency of the inner rings of Saturn. Before his eager eyes, mist-covered holes opened in the sun's chromosphere, red spots glowed on Jupiter, the tail of a comet pulsed with light. The skies, out of reach, must have seemed endlessly complex.

The Harvard Observatory commissioned a series of astronomical drawings by Trouvelot, which drew favorable reviews in the *New York Times*. A critic wrote, "In one of these drawings there is a fork of flame very nearly 100,000 miles in height, and it is surrounded with cascades of fire resembling—though on an immense scale—the fantastic play of the flames which leap from the huge furnaces of the iron districts."

One day in May 1883, while the gypsy moths in Medford

consumed his former neighbors' fruit trees, Trouvelot perched on Caroline Island, an isolated reef in the middle of the South Pacific, waiting for a solar eclipse. In a break between two storms, he watched with other scientists from the United States and France as the sun ducked out of sight and the ocean was lit by stars. For the five minutes and twenty-three seconds of darkness, he scanned for planets orbiting Mercury but found nothing.

⚹ ⚹

AT the moment when I get in the car and turn to go, releasing Myrtle Street to its history, European gypsy moths are spread throughout the eastern half of the country. An Asian cousin recently arrived in California by boat, and both species have turned up in evergreen-spotted states, where trees are big business and don't recover from losing their needles. The European gypsy moths have been named and renamed over the years—no longer viewed in the same genus with *Bombyx mori*, the silkworm, they are now dubbed *Lymantria dispar*—but the destructive effects are the same. In 1981 they defoliated twelve million acres, an area the size of Massachusetts, with Vermont thrown in for good measure. Managers spray gallons of Dimilin, Gypcheck, and other new insecticides. Their tricks also include luring males to their deaths with artificial sex pheromone. Since the first forays at the turn of the century, scientists have introduced more than one hundred exotic predators, hoping to control the outbreaks. In 1989 one of these paid off.

Early that summer researchers in a Connecticut forest found gypsy moth larvae hanging, head down, from the bark of oak, maple, and birch, trees they normally would be consuming. Those longest dead were blackened, their sides caved in. Examinations revealed a fungus from Japan that infected the larvae and used the corpses to grow and transfer spores. The fungus, *Entomophaga maimaiga*, turned up in gypsy moths from Vermont to

Pennsylvania, dispatching them with an effectiveness that would have gladdened the hearts of Trouvelot's neighbors. Was this a mutation of Clinton's strain, dormant for eighty years, or a recent introduction, unnoticed until it flourished in the rainy and humid spring of 1989? No one is sure, but either way, the fungus offers new ammunition for the battle in the woods and provides another plot twist in the century-long drama.

More traditional methods of control linger as well, and on hot summer evenings in the East, families still go seeking gypsy moth egg clusters, torch in hand. During an outbreak not long ago, while caterpillars moved like a black carpet over six states, the plague became a game. Kids perfected the art of stomping on the caterpillars at just the right angle so the guts would spray all over their friends.

⋇ ⋇

PACING the streets of Medford, Trouvelot must have walked around in a daze half the time. With knowledge not only of the interstellar dramas crashing overhead but of the minute worlds writhing through the microscope, he must have tried to grasp the complexity visible at every scale. Strange, though, that with the sun shrugging off million-mile-long licks of flame like so much comet dust, he counted on the microscopic world to follow a regular plan. While he watched an eclipse make day into night as scientists helpless to exert control jotted notes, he thought that an insect might easily be harnessed to industry, that the natural world would follow his dreamed-up rules. It's a common enough assumption, and many stake their fortunes on it. But Trouvelot? With all that he knew, with all he'd seen, he surely might have suspected that control of the natural world might elude him.

CHAPTER SEVEN

FLUSH WITH SUCCESS

THERE, rustling in the brush. A glint of red, a gleam of blue. Hard to separate from the snare of branches, difficult to pick from the speckled shade, until one crew member swept it up in his arms, feeling the heart race against his chest, the wings beat at his face. For a brief moment he couldn't distinguish his fingers from the feathers, one pulse from the other, until he released the bird in the ship and caught his breath. Then along with Medea, along with the Golden Fleece, along with the knowledge that an entire army of men, grown from dragon's teeth, can be slain with a rock, Jason and the Argonauts carried the pheasant back to Corinth.

This legend of how the pheasant got to Greece lives on in the bird's Latin name, *Phasianus colchicus*. As the Argonauts sought the Golden Fleece, they rowed the *Argo* up the Phasis River to the kingdom of Colchis. The river, now dubbed the Rioni, still flows down from the Caucasus Mountains, through Georgia, into the Black Sea, and pheasants live along the banks. When the Romans conquered the Greeks and pushed northward, they brought the pheasant with them to England and the rest of Europe.

Hunting the black-necked pheasant, the bird with such an illustrious introduction story, went on to become the sport of kings. But there was always a suspicion that something even more fabulous lurked in the Orient. In the thirteenth century Marco Polo returned from his travels in China with tales of a wondrous bird twice the size of European pheasants, flaunting a tail ten feet long. He didn't bring any back with him, so these hints festered in the mind, a clue still to be investigated.

⁂   ⁂

IN July 1880 Judge Owen Nickerson Denny and his wife Gertrude threw an Independence Day party in Shanghai. In the twilight lanterns traced the path to the consul general's house, the Dennys' home. Music slipped out the windows into the garden. Dancers panted and laughed and cooled off with gulps of ginger beer. More than four hundred guests clinked glasses as Owen Denny toasted "To the day we celebrate."

In the pulse of the party, Denny wouldn't have been able to escape business. As China was shrugging off isolationism and engaging in trade with the West, it sought advisers to help it master Western communication. Denny, who had seen starved bodies floating in the river and people eating tree leaves to stave off hunger, urged China to make better use of its resources, to mine for copper and lead, to build railroads to transport food from one place to another. He also had the unenviable task of explaining to Chinese officials why the U.S. government was pursuing a bill that would prohibit Chinese immigrants from ever becoming citizens and why Chinese men were being lynched and shot in the streets in San Francisco and Seattle. Fortunately he and Li Hungchang, the viceroy, developed a good relationship, and Denny was a popular official.

If, in a break from hosting, Denny and his wife glanced over at each other, she might have admired his curly dark hair and the familiar blue eyes that tipped down at the corners, giving him the stricken expression of one constantly receiving shocking bad news. He might have appreciated the frank, round face that had drawn him to her in their home state of Oregon, though now they were so many miles away and she had put away her banjo and adopted the dress of a diplomat's wife, costumes of heavy satin and brocade appropriate for hosting heads of state and sipping lemonade with royalty.

The Dennys were drawn together by a mythology of their own. Both had come to Oregon with pioneer families, and the directive to homesteaders who would claim their acreage was always the same: Improve the land. A plot was only as good as the house and the irrigation and the domestic animals on it. All around him in China Denny saw species that would make good improvements to his home state. There were flat peach trees in the orchard, bamboo and pheasants in the garden.

The birds wandered wild through fields and vegetable plots, and Denny bought them from Chinese farmers. Their beauty stirred his imagination: "On one occasion I had in my enclosure a large number of extraordinarily handsome birds, and while admiring them I thought, What would I not give to be able to turn the entire lot adrift in Oregon?"

Even with his diplomatic burden, the judge found time for leisure activities. In an empty hour he stalked the jeweled prey outside town. The gleaming feathers disappeared into dust and branches, then reappeared in an open meadow. Gone again. Finally the pheasant flushed out of a tangled garden, where it had been pecking at vegetables. Denny took aim and, in the middle of the squawking and flying feathers, heard a voice: Don't shoot. The Chinese owner of the plot appeared. But they're eating your vegetables, argued Denny. They eat the insects too, the farmer explained. They do more good than harm. At least that was how Denny told the tale back in Oregon to the farmers who watched imported pheasants patrol their hard-tilled fields, claiming ownership.

Denny ended up giving quite a bit of both time and money to see his idea realized. In early 1881 he packed about sixty pheasants along with Mongolian sand grouse and Chefoo partridges on the *Otago*, headed for the United States. But the birds panicked and bashed themselves against the sides of their cages as they were

transferred from the dark hold of the ship docking in Port Townsend to open containers on boats and trains bound for Portland. By the time a friend of Denny's released them near the Columbia River, too few were left alive and too many of those were males to establish a colony. The attempt cost three hundred dollars, a hefty chunk of Denny's monthly salary.

The next winter, 1881–82, Denny tried again, taking extra care. He measured the airy dimensions of the aviary he'd had built in the hold of the *Isle of Bute*, enclosed by bamboo poles. He ordered sand for the ground and grain for food. He asked his brother to pick up the birds in Portland when they arrived in March and release them on Peterson's Butte, behind the Denny homestead near the town of Lebanon. He convinced the Oregon legislature to outlaw hunting the pheasants for five years so the population could grow. Then he waited for news.

From his post in Shanghai that was all he could do besides dream about the release. If they made it through the winter and the boat trip, the males would start marking out their territory in spring, the largest claiming the prime spots where open field met concealing brush. Courtship displays would follow, the cock spreading a wing in front of the female, ducking behind it, like a fan dance, and finally feeding her choice bits of grain. He might do this once, twice, or three times, building a harem. The chicks, when they hatched, striped and fluffy with long pink legs, would learn to pick at and catch insects of the Pacific Northwest rather than China. In the maze of the foreign city, fingering a porcelain vase or drafting a business letter, Denny could only remember the open country of the Willamette Valley where his family had first staked their homestead claim when he was a young boy, near where the Santiam River met the Willamette and the first bumps appeared that eventually swelled into the Cascades. He could only picture boys running through the woods, breathing in cedar and the soupy smell of river, whipping bushes with sticks, until out of one of them a live fireworks explosion burst, a blast of feathers and croaking, a pheasant. They would be left, nerves wakened, hearts racing, wondering, "What was that?"

Nothing looks more exotic than a male ring-necked pheasant. His back shines gold and green and rust, a harvest of color, and his head is even wilder. Red wattles surround the eye, a pink blush glows on the throat, more gold and green mark the neck, and two tufts stick up like ears. He has a long sword of a tail. A band of white encircles the neck like a collar. On the other hand, the female is all soft brown flecked with black. She has a shorter tail, though no less pointed. She's evolved to blend in, while her mate evolved to stand out. But underneath all the splash and dash, the spangles and iridescence, the ringneck is vaguely reminiscent of a big chicken.

The similarity is more than superficial. Along with grouse, turkey, ptarmigan, and quail, pheasants and domestic chickens share the order Galliformes, which means "fowllike birds" but might was well be translated as "birds that are tasty." They generally have round wings, four toes, large clutches of eggs, and squat, plump bodies. The black-necked pheasant of the Caucasus Mountains and the ring-necked pheasant of Shanghai are two varieties of the true pheasant, *Phasianus colchicus colchicus* (the black-necked pheasant) and *Phasianus colchichus torquatus* (the ring-necked pheasant). The long-tailed bird glimpsed by Marco Polo was probably yet another type, *Syrmaticus reevesii*, or Reeve's pheasant. Pheasants and chickens share vocal traits as well. As startled Oregonians would soon discover, some ringneck cocks crow at dawn.

⊀   ⊀

AFTER Owen's brother John had lugged the pheasants up to Peterson's Butte, hidden behind a tree, and opened the cage, residents of Linn County watched for a long tail darting across the road, the flicker of iridescence in the underbrush. The fact that pheasants lurked in the stubble along field edges was no secret. On cold nights the birds wandered into the barnyard to pick at chicken feed or crept into the henhouse during a snowstorm. The law didn't allow hunters to kill any yet, but it couldn't prevent anticipation. Even the local newspaper cooperated, coyly suggesting that reporters knew where the pheasants had been released but wouldn't tell, merely whetting readers' appetites with the note "Ten years from now, no swell dinner will be complete without them, thanks to Judge Denny of Shanghai."

When Denny returned to Oregon from China in late 1884, he brought more ringnecks, as well as copper pheasants with feathers tipped with burnished brass, golden pheasants with blond pompadours, Japanese green pheasants, chests and backs splashed with emerald. Released on Protection Island in Washington State, only the ringnecks persisted.

In the summer of 1886 the Dennys were back in the Far East, unpacking dishes, repotting plants brought from Shanghai, and building shelves in their new home in Korea, where Denny had been appointed adviser to the king. In the cool evenings, after stifling hot afternoons, Gertrude Denny sat on the veranda and looked out over her freshly sodded lawn. Back in Oregon the ringnecks also settled in. The landscape that they found themselves in was, perhaps more at that moment in history than at any other, pheasant heaven. Abandoned wheat fields grew into thickets rich with grain as the wheat industry moved to eastern parts of the state and Willamette Valley farmers cast about for new ways to use their fertile soil. With this ample supply of food and habitat, the birds spread out, dodging hawks and coyotes, tucking clutches of eggs into grass pockets, reaching and then crossing the broad-shouldered gray-blue Columbia, shoving its way to the sea. In 1887 the Oregon legislature extended the pheasants' legal protection for five more years.

Hunting lore has it that on the first day of the first pheasant season in 1892, gunners killed fifty thousand ringnecks. As their dogs picked up the wild scent, the hunters pursued the cocks through the underbrush, the longtails flew out over the fields, and the victors pulled the triggers and posed for photographs standing above the limp bodies lined up in rows, Oregonians confirmed the reports: Pheasants were good hunting.

The ringneck has all the personality of an animal designed to provide prime sport. Sneaky and quick, the pheasant poses enough of a challenge to make it valuable quarry. In some mysterious way it can fold up all those bright gleams and disappear into a bush, hiding until the moment when it flushes and bolts over the fields. The long streaks of wing and tail blur against the sky until a shooter cuts short the surge. As one naturalist summed it up, "They fly swiftly, run fast, seem intuitively to know a gun at sight and are remarkably tenacious of life." He did add that "the closest observers contend that they are slowly but surely crowding our native birds."

As hunters from Washington, California, and even New York returned home from Oregon jaunts, they told tales of the pleasures of pheasant hunting. One successful pheasant breeder in Oregon commented in 1899, "It is not too much to say that visiting marksmen who possess the true instinct of the sportsman, and have enjoyed a day's shooting in this valley, will never be satisfied until their own preserves and fields are well stocked with these festive game birds." She was right. In no time other states demanded pheasants of their own. Individuals, hunting clubs, and state fish and game departments gathered eggs and shipped adult birds across country. Enterprising souls began raising the birds in backyard coops, sliding pheasant eggs under domestic hens. Oregon pheasants arrived in Sturgis, South Dakota, in 1891. Montana had brought in pheasants by 1895. Minnesota sought ringnecks in 1905. At times, with closed hunting seasons the first few years, the birds dug in and established themselves along the field edges and in the underbrush. In other places they vanished. Currently nineteen states boast pheasants.

The introduced game birds served two purposes: They filled the gaps left by the rapidly disappearing native species, and they provided an exotic hunting experience close to home, a sort of democratization of the tradition in which wealthy sportsmen would roam the world searching for odd and fierce things to shoot. One sportsman wrote for *Harper's Weekly* in 1897 about taking the train from New York City down to Long Island and hunting exotic birds that had been introduced to a small island off the coast. He reported, "This is the sport of civilization, and the men of Robin's Island have solved the difficult problem by bringing the free open air, the wild life of the woods and sea, with sport as thorough of its kind as can be found anywhere, into such close proximity to work and home and duty that the whole thing can be done in less than thirty-six hours." Not everyone could go to Shanghai.

The success of the pheasant prompted the release of other exotic birds for hunting. The chukar partridge, a red-legged bird of the Himalayas, arrived in Illinois from India in 1893. While thousands of these birds didn't take, including an import of 84,414 from Nepal set free in Minnesota, the chukar still roams dry hillsides through Nevada, Idaho, and other states west of the Rocky Mountains. State agencies, gun clubs, and individuals sought out and set free Eurasian quail, black partridge, black grouse, and Hungarian partridge to scurry into the brush, rear large broods, and appease the sport hunter's appetite.

⚹ ⚹

WHILE homesteaders may have looked on in amazement and hunters crowed with glee as the gleaming ringnecks wandered into the cornfields, it was not the first time people witnessed pheasants in the New World, or thought they did. Early observers noted something that looked like a pheasant running around the underbrush in Massachusetts, New York, and Pennsylvania. In

1637 Thomas Morton made the comparison overt: "There are a kind of fowles which are commonly called Pheysants but whether they be Pheysants or no, I will not take it upon me to determine. They are in forme like our Pheysant henne of England. Both the male and female are alike: but they are rough footed and have stareing featehrs about the head and neck; the body is as big as the pheysant henne of England; and are excellent white flesh and delicate white meate yet we seldom bestowe a shoote at them."

Eventually colonists realized that the rough-footed bird with the pheasant silhouette was not the genuine article, and they set about bringing it over. The marquis de Lafayette sent pheasants to George Washington, but Washington was more interested in the possibilities of the donkeys that accompanied them. He hoped they would populate the country. Richard Bache, Benjamin Franklin's son-in-law, also imported some pheasants to New Jersey, but they didn't take. Neither did those brought to New Hampshire in 1790 by the state governor.

The bird described as a pheasant by Morton was likely the heath hen, a bird facing grave danger just as the pheasant population was beginning to take off. The heath hen had roamed throughout the East, through blueberry patches and grasslands, providing meals for early settlers, but not very much fun. Elisha J. Lewis, author of *The American Sportsman*, wrote of the heath hen: "So numerous were they a short time since in the barrens of Kentucky, and so contemptible were they as game birds, that few huntsmen would deign to waste powder on them." Though they didn't satisfy the sport hunters, the heath hen attracted enough bullets from market hunters to assure its demise. By 1882, as the ring-necked pheasants were tending their newly hatched chicks in the Willamette Valley, the last population of any size of heath hen pecked in the grass on Martha's Vineyard.

The attitude toward native game birds, dismissing them as not posing enough of a challenge, was common. Many species had been dubbed fool hens because of their apparent willingness to sit still and be shot at. The sooty grouse, the white-tailed ptarmigan, the sage grouse, the Columbia sharp-tailed grouse, the Oregon ruffed grouse, the mountain partridge, and the valley partridge were nothing to the ringneck and its wily ways. At first the hunters were ready to kick out the more humble native relatives to let the ring-necked pheasant strut his stuff. In 1893 a man who wanted to stock an island with ringnecks, fearful that the native quail might interfere, planned to remove them. S. H. Greene underscored the point in a letter to *Forest and Stream*, writing of the ringneck, "[O]ur local sportsmen are about unanimous in their opinion that where he gets a start, the aboriginal pheasant and grouse 'must go.' " Of the sooty grouse, a notoriously tame bird that was vanishing from Oregon, William Shaw recalled, "Men, still young today, tell you that in their boyhood they have actually killed the fledglings of these birds with sticks as they fed about shocks of wheat. Others tell of shooting whole flocks, picking them off, one by one, from a fence or tree branch, down to the last bird. They were not meant to withstand civilized progression."

Whether beating chicks to death or killing off an entire population is a sign of civilized progression is up for debate. Either way, many of these sportsmen believed that Darwin and his theory of natural selection were on their side and that just as the expansion of Europeans was inevitable, so was the replacement of species that couldn't coexist with them by species that thrived on human contact. But others felt concern and anxiety, expressed in the pages of magazines like *Harper's* and *Forest and Stream* as the heath hen numbers continued to plummet. These writers envisioned themselves in many ways as the protectors of game, warring against adversaries determined to wipe out species. At the one end of the scale was the ruthless market hunter, who shot for profit, and sold the plumes for hats or the carcasses for restaurant tables. At the other end was the lowly pothunter, who went out looking for food and ignored the rules and traditions that made hunting a sport or art rather than an occupation. A contemporary comparison might be the feeling that some fly-fishermen have for ocean trawlers raking in fish with gill nets, on the one hand, and bait fishermen with their cans of worms, on the other. But despite individual efforts, along with those of government, setting aside reserves and levying fines on poachers, the heath hen was quickly and determinedly slipping from their grasp. In the early 1890s, as the first hunting season for ring-necked pheasants opened, only about a hundred heath hens remained. Replacement with exotics may have seemed not only desirable from a sporting standpoint but necessary to keep the woods from standing empty.

Recently problems with this decision have become apparent. While heath hens crossed paths with ringnecks only briefly, if at all, the hen's western relative, the greater prairie chicken, inhabits land the ringneck loves. In contrast with the flashy import, nothing looks more like a creature of the American tallgrass prairie than the greater prairie chicken, another of the Galliformes. The westernmost variety of its species, the greater prairie chicken (*Tympanuchus cupido pinnatus*), ranges in the western plains, while the heath hen (*Tympanuchus cupido cupido*) lived in the Northeast, and Attwater's prairie chicken (*Tympanuchus cupido attwateri*) dominates the Southwest. Bands of brown and white cover its chest and back and wings, evoking dust and twigs and the hollow stems of dry grass. The mating strategy of prairie chickens is dependent on the open spaces where they live. During mating season, males display and dance simultaneously at lekking sites or booming grounds. In an effort to attract females, they inflate orange teardrop-shaped sacs on their throats, and neck feathers stand above their heads like a ruff or crown. They bow, stomp their feet, and emit eerie hooting cries. More experienced males claim the prime dancing spots near the center of the circle and mate with more females as a result. After the performance the females leave to raise the chicks on their own. The booming grounds are traditional, and populations of prairie chickens return to them year after year.

But the booming grounds are giving way to fields, houses, and businesses, and the newly arrived ringnecks eye the remaining patches as they carve out their territory in the grasses. Ringnecks can hybridize with native species, including the sooty grouse, blue grouse, and ruffed grouse, muddying genetic waters. Female ringnecks have been known to slip an egg or two in a prairie chicken's nest, resting assured that that egg will hatch several days before the prairie chicken's and monopolize its adoptive parent's attention. Male ringnecks will attack and chase prairie chickens for up to a mile. At times this aggression causes prairie chickens to abandon their booming grounds and move on, though often, with fields and roads on every side, there isn't much place to move on to. The ringnecks may be just taking advantage of situations created by a loss of habitat, but the fact remains that the greater prairie chicken is threatened, Attwater's prairie chicken, the southwestern variety, is critically endangered, with only a handful left, and the heath hen is extinct. The decline of the prairie chicken has become a burden for another generation.

※ ※

ON I-5, south of Portland, a sign welcomes visitors to Linn County, "the grass seed capital of the world," and at the entrance to Lebanon itself another message announces "the city that friendliness built." Peterson's Butte itself rises, many-humped, right outside town, visible from Main Street. The road leading up to the north side of the butte is called Denny School Road. Lined with blackberry brambles, it twists through a landscape of fields and backyards, old white farmhouses with porches, and cluttered trailers. In early summer green pulses through everything. Clumps of Scotch broom wave their yellow tentacles in the air.

Killdeer dart out of the pastures, sharp wings over the asphalt. Blackbirds peck at the occasional road-killed skunk, and sheep graze behind wire fences. At the butte's base a sheep scratches itself on a metal trough underneath a feathery cedar. There's no commemorative plaque, but it's not hard to imagine the first pheasants settling in without much difficulty.

Rarely has an exotic species been so warmly embraced, so quickly woven into the fabric of America's idea of itself. South Dakota, blessed with millions of pheasants, declared it the state bird. Ringnecks stare out from postcard stamps and flutter across vanity checks. When pheasant populations dipped at mid-century as a result of changes in agriculture that left fewer patches of brush at field edges, sportsmen rallied to help the ringnecks recover (by this time the heath hen was long gone, the last one having died in 1932). Today pheasants are farmed and trucked into prime habitat by the thousands, just before hunting season. On game farms the sportsman can choose his bird, watch it be released on a scant few acres, then chase it down. Living on the edge of human settlement, flirting with domestication, the pheasants still remain just on the wild side of the line.

While importers of many exotics are cursed and reviled as all their good intentions unravel, after her husband died in 1900, Gertrude Denny received a monthly pension from an Oregon sporting club in gratitude. In a gesture similar to that of the Greeks who encoded the pheasant's introduction story into its name, the press association voted to refer to the bird as the Denny pheasant, but it never really caught on.

CHAPTER EIGHT

TROUT DIPLOMACY

On cold September days, when snow slants into the water, the brown trout surge upstream, a current of muscle pushing against the waves. Their colors echo the river—bronze, greened copper, flecks of gold—but their motion contradicts the flow. Grown fat on mayflies, stone flies, whatever they can pick off the surface or snag in their jaws, they head to gravelly riverbanks to lay and fertilize eggs, tending to the next generation of this phenomenally successful fish. The anglers, numb-fingered on the shore, stamp their feet, cling to their rods, and anticipate a good fight, looking forward to pitting their skill against these wily and tough trout.

A fisherman friend tells me, "Four out of five of my best fish stories are about brown trout."

And I say, "Listen to this."

※　※

THE fish grew thinner, more transparent, every day. Newly hatched and no bigger than a child's fingernail, each was only an eye and a sliver of body with a yolk sac still attached. They clustered in their water-filled milk cans—a hundred thousand shad in all. Fred Mather, the assistant to the United States Fish Commission, the man in charge of ensuring the fish landed safely in Germany, tended them carefully. Every hour, night and day, he and his coworker changed their water, pouring it from pail to pail to increase the oxygen levels. In an effort to keep the water clean, Mather skimmed off the dead, so the survivors would have a better chance. The first day he scooped up five hundred. Two days later he removed a thousand. One long night fog wrapped the ship like a gem in cotton, and the heat made it hard to think straight. The next morning three thousand dead fish floated on the surface. While the yolk sacs still clung to them, they struggled on, sucking the nourishing fats inside. But when, in the course of development, the sacs disappeared and they had to forage for food, they dropped quickly. No one knew what baby shad ate, so Mather's goal was to get them into a stream where they could fend for themselves before they starved. Young fish had made it across the United States—a trip of seven days on the train—but a steamer trip from New Jersey to Germany in 1874 took ten. Mather grew nervous as he watched the tiny fish fade and offered them chunks of raw meat. He changed their water every half hour instead of every hour. Then, one morning, he peered in the milk cans, and all the fish were dead.

When he recovered from the trip and caught up on his sleep, Mather converted his discouragement to productivity. He thought over the journey, calculated its mistakes, and started experimenting. The failure had been a failure of technology, and both Mather and the Fish Commission hoped the right gadgets would make transporting fish across the ocean as easy as scooping them out of a pond. First he worked on a technique for hatching shad mid-voyage, so they could reach the other side before starving. Fish kept dying in his hatching can, though. In one experiment the mesh where the eggs rested proved too big and the eggs fell through. In another he used water kept in an old whiskey barrel and thought the trace of alcohol might have done them in. In a third, Mather speculated, the tiny fish were harmed by a decomposing whale in the basement where he made his tests. Finally Mather's assistant suggested he modify his design. The new hatcher, a funnel with a hose at the bottom attached to a tap, included a mesh basket near the top where the eggs sat as the water rushed over them and out a spigot on the funnel's upper rim. The moving water stirred the eggs, mimicking the push of a current. The two men tested it on the porch of a Pennsylvania hotel and released the hatched shad, frisky and healthy, into a nearby river. Additional tinkering produced a box that kept salmon eggs cool en route so they could hatch on a European shore.

Technology improved all the time, and within six years of Mather's first attempt, Americans transported fish to Germany with ease, offering many species the Old World had never seen before. The Germans loved the shad, the California salmon, the silvery rainbow trout. In gratitude the German Society of Fish Breeders sent a letter to the U.S. Fish Commission, probably written by the enthusiastic head, Herr von Behr. "We should hail the day, dear sir, when we might be permitted to offer you, for the benefit of American rivers or lakes, any inhabitants of our waters unknown beyond the ocean," the letter read. The Americans took him up on it.

꙳ ꙳

WHILE Romans filled pools with exotic varieties of fish to satisfy their appetites, and ancient Chinese strained fertilized eggs out of creek bottoms and carried them from place to place, fish culture was just hatching as a government enterprise in the United States in the second half of the nineteenth century. The abundance of wildlife noticed by Marc Lescarbot and other early visitors was waning. In the Great Lakes, fishermen complained that despite more boats and better nets, they brought in fewer fish each year. On the East Coast waters that used to be thick with cod and pulsing with mackerel now glinted empty. Dams, overfishing, and logging were taking their toll. In response, hobbyists of the 1860s experimented with fish culture in backyard ponds. They protected young fish from predators and imported a variety or two. After returning from fighting for the Union in the Civil War, Mather himself dabbled in the business, buying a farm near Albany to raise and sell the eggs and fry of alewives and yellow perch.

In 1871 the U.S. government launched the Fish Commission, headed by Spencer Baird, assistant secretary of the Smithsonian Institution, to investigate the disappearing fish. While its main purpose was raising and stocking local varieties, ensuring that the American people would be fed, the commission also experimented with transplanting successful species from one area to another. Trout had been shipped as far as from England to Tasmania in 1864, but that was only the beginning. Everyone could use a better fish.

With the transcontinental railroad cabling one coast to the other in 1869, and shipping methods improving all the time, fish could be carried farther and faster than before. Ponds and lakes, rather than being viewed as complex ecosystems, were treated as outdoor aquariums waiting to be filled. Once walled in by the boundaries of their pools or watersheds, now eastern fish traveled west, western fish journeyed east, English fish came to the United States, American fish swam through streams in Australia. Leather carp were whisked from Washington, D.C., to Scotland; Lake Michigan whitefish found themselves in New Zealand. German carp settled in Alabama, while Hawaiian mullet traveled to the mainland. Between 1874 and 1880 the Fish Commission transported more than four million California salmon to foreign countries, including Canada, England, France, Holland, Germany, New Zealand, Australia, and the Sandwich Islands.

These early attempts, like Mather's, were a race against death. On one shipment of live lobsters from Maine to California, the fish culturists juggled sponges, damp straw, salt water, and blocks of ice in an effort to keep their charges cool, damp, and alive. Then the lid of one aquarium slid shut, suffocating all the crustaceans inside. As the trip continued, still more died. The fish culturists dropped off one of their few remaining pairs of lobsters in Utah to breed in the Great Salt Lake. By the time the lobsters reached California, only four lingered to be released from a wharf in Oakland. Another California-bound shipment, an aquarium train car filled with oysters, black bass, lobsters, shad, catfish, walleye pike, bullheads, yellow perch, and trout was rolling through the plains when disaster struck. On a bridge spanning the Elkhorn River in Nebraska, the aquarium car broke away from the rest of the train and tumbled into the water. Fish tanks, ice, lobsters, and eels catapulted through the air. One man was crushed to death. The fish culturist in charge moaned, "No care or labor had been spared bringing the fish to this point, and now, almost on the verge of success, everything was lost." However, the head fish commissioner was more optimistic. While the saltwater fish promptly expired, he admitted, maybe some freshwater varieties had established themselves in the Elkhorn.

⚹ ⚹

IN the hierarchy of pisciculturists in the United States, Mather was not that high. Baird's name was on everyone's lips, employees like Livingston Stone released more fingerlings, and Seth Green's inventions earned more kudos. But no one could top Mather for sheer enthusiasm, an infectious love for fish and their propagation. In an article in the first issue of *Forest and Stream*, he mocked those who thought of fish culturists as wizards and fish breeding as "a strange, new business, that has, like the genie of Arabian Nights, sprung suddenly from almost nothing into vast proportions." He went on to predict a magical transformation in the country's lakes: "As the Hudson River has furnished those living near it with tons of cheap and wholesome food, so can each of these beautiful lakes, which are now only so much waste space taken from agriculture, be made to yield a larger return, acre for acre, than the best grazing lands in the state." While many of *Forest and Stream*'s readers wanted to introduce good sport fish, Mather stressed that he didn't care how much fun a fish was, the mission of fish culture was to provide food. His opinion matched that of Herr von Behr, head of the German Society of Fish Breeders and Mather's friend. While Behr was a German baron and Mather an American country boy turned miner, soldier, and bureaucrat, if conversation ever faltered, they could find common ground in the belief that in fish culture lay the future of the world.

By 1880 Mather had done enough right to be selected as the Fish Commission representative for angling and fish culture technology to the German Fisheries Exhibition in Berlin. International exhibitions were a fixture of the second half of the nineteenth century. Ever since the London Exhibition in 1851, when the Crystal Palace and its contents blazed with England's glories, country after country built new, shiny pavilions and sent out invitations for festivals of their own. These celebrations of technology and progress, with a little subtle jostling for cultural superiority, often fractured into events focused on a single industry: agriculture, transportation, art. Part spectacle, part business, they resembled huge marketplaces with ideas and designs, and sometimes species, as the currency. By the opening of the exhibition Mather's initial struggles with all those dead shad were scarcely a memory, and the United States and Germany swapped fish the way collectors swap baseball cards.

Even for the casual observer, the fishery show overflowed with pleasures. Crowds reaching twenty-four thousand on some days surged through the halls. Past the central rotunda, where a statue of the god of the sea presided over the Grotto of Neptune, tables displayed innovations from Japan, India, and South America, as well as most European countries, including Russia. China offered an exhibit on cormorant fishing, where fishermen placed rings around the necks of the birds that prevented them from swallowing and released them to dive for fish. Germany set up models of fish ladders, which allowed fish making their way upstream to cross over dams. Smells of smoked halibut, sardines in mustard, and pickled soft clams hung in the air. Vials of whale, codfish, and sea elephant oil glowed dully, offering viscous remedies. Those tempted by luxury could covet abalone shells and pearls or run their fingers through sea otter furs. Professionals, on the other hand, could compare technology, discuss international fishing laws, trade maps of the ocean floor, and scout for desirable fish to bring home and cultivate.

For someone of Mather's bent, the exposition promised a garden of delights. He sampled Russian fish delicacies and pronounced them tasty. He scoffed at scientists' claims that they could reliably determine the sex of an eel without dissection. He admired a display of English reels and fishing lines. He glanced at a live brown trout swimming in a tank and dismissed it as not as pretty as the American brook trout. He fingered amber cuff buttons and reflected on the lives of ancient insects trapped thousands of years ago in the blood of a tree.

As in many of the fishery exhibitions of the time, America starred. Whale harpoon guns, watercolors of Alaskan seal fishery, dories rigged up for a hard day's fishing, and a model of the *Fish Hawk*, a ship specifically designed to transport roe and fry safely overseas, awed visitors. The United States collected gold medals as if they were clamshells on the seashore and even picked up the first prize of honor, a crystal and gold tower of serving platters. In the center, a boy holding a dolphin on a spear perched on a globe, ruling the waters. Mather gathered gold medals for his fish-hatching invention and his apparatus for carrying eggs across the ocean and a bronze medal for machinery that kept water aerated, saving the labor of pouring it from bucket to bucket every hour. One German official commented, "[T]he American department was a complete fishery exposition in itself."

But for a break from the sweaty handshakes and elbowing

gawkers, Mather slipped away to a different realm of diplomacy. He went fishing. After the load of professional responsibility, he must have expanded in the quiet. No hearty business chatter, no machinery, just the push of water against stone and the familiar rhythms of the sport he'd loved since childhood. Angling in a stream in the Black Forest, breathing in the pine, he hooked a fish. The trout in his hands glowed yellow underneath, with black and red spots on the upper body, loose and large like the first drops of rain. Around each spot a pale halo radiated outward. He had hooked a *Salmo trutta*, a brown trout, and something about this wild fighter made him forget the fish he'd dismissed at the exhibition. Maybe the trout struggled, dragging his line through the water, thrashing as no brook trout ever had, and the rush of adrenaline clouded Mather's memory. Maybe it was huge, and he thought of how many people it could feed. Mather didn't record the trout's dimensions, but for whatever reason, he wanted to see it flashing in the currents and hiding in dark eddies along the Hudson. Before leaving for the United States, he mentioned it to Herr von Behr, perhaps reminding him of his offer.

꒝ ꒝

BACK in Long Island, waiting for his brown trout, Mather headed the new hatchery at Cold Spring Harbor, a converted whaling station. In one pond wood ducks paddled around, shaking their flashy green heads. Standing beside long wooden basins, men in vests and bowler hats plucked the dead fish eggs from live ones. Particularly designed to breed saltwater creatures, the hatchery planned oyster culture and a pool that would fill with fresh seawater at every tide. In his spare time Mather wrote fishing columns for *Forest and Stream*.

Then, as inspectors fussed over Cold Spring Harbor and discussed expansion, Behr proved good to his word. The steamer *Werra* brought eighty thousand brown trout eggs and unloaded them on February 28, 1883. Mather considered them a personal gift and kept many at Cold Spring Harbor. He also distributed some to hatcheries in Caledonia, New York, and Northville, Michigan. Many arrived partially hatched and dried out, and few or none made it into the rivers. Never daunted, Behr followed up with a more successful shipment a year later. In addition to these German gifts, the editor of the *London Fishing Gazette* gave ten thousand brown trout eggs to the New York Fishing Commission. Of Behr's second shipment of sixteen thousand large and fifty-four thousand small, Mather sent half the large and twenty-one thousand of the small to E. G. Blackford, fish commissioner for the state of New York. He also sent three thousand large and ten thousand small to Caledonia, a thousand large and four thousand small to Northville, and two thousand large and nine thousand small to Central Station in Washington, D.C. Others he kept to raise in the Cold Spring Harbor ponds.

Mather found the trout beautiful. When visitors wandered through the hatchery, he scooped the fish out of the shallow wood-walled ponds. After displaying the scarlet speckles, Mather released the fish again, but the morning after each performance, dead trout littered the ground outside the pond walls. Finally, when only fifty fish remained from Behr's first shipment, Mather noted, in dismay, "This fish seems given to this form of suicide. . . . I had no intimation of this habit from any of my European correspondents, and the fish differs in this respect from our own trout, which readily accepts capture and transfer." Though most died, Mather dubbed one hefty survivor Herr von Behr.

Meanwhile, at the fish hatchery in Northville, Michigan, in April 1884, manager Frank Clark didn't fiddle with his stock but released it, dropping his young brown trout, no longer than peanuts, into a branch of the Pere Marquette River. A dozen could be cupped in a palm.

Brown trout must be flexible and forgiving, for they took to the foreign ripples, the streams where none of their ancestors had

flourished, and consumed the alien insects with relish. For the first two years they sipped down stone flies, caddis flies, mayflies, bumblebees, and ants. They lurked behind rocks and let the water drag drowned nymphs to their gullets. They put on weight. In the third year the females scooped a nest in the gravel, tested its depth with their fins, maybe scooped some more. The males hovered nearby, fighting off others that approached the females' work. When they dug deep enough, the females deposited eggs alongside males releasing milt, creating a cloud of underwater smoke. Before moving on, the females flicked gravel over their nests, a layer of protection.

In the chambers between pebbles, the eggs developed, each gaining a dark center, an eye. Water rushed through the bed, stirring the eggs and bringing oxygen, and soon tiny fish hatched. Large yolk sacs slowed their movements and provided sustenance. Still cradled by the rock nest, the alevins absorbed the rest of the yolk, their mothers' last gift, and then wriggled into the current. They inched toward food, darting after zooplankton, then mayfly nymphs, then caddis fly larvae, whatever could fit in their mouths. A kingfisher may have swooped down to pluck a fingerling out of the spray. An otter crunched another. The older fish with larger stomachs to fill switched to eating smaller fish, preferring brook trout to brown and crayfish, if they could find them. In no time they bossed the river. Four years after the first release, as the eggs from the year before swelled in the nests, a man angling in a creek near the Northville hatchery pulled out a brown trout measuring twenty-two inches, long as an arm, shoulder to wrist.

꙳   ꙳

MATHER didn't live to see the full popularity of the brown trout; he died on Valentine's Day 1900, just after assuring *Forest and Stream* that he had another article in the mail. By then brown trout swam in thirty-eight states, from New York to Colorado. They eventually were hauled out of streams in the Great Smokies and the Missouri River. While Mather insisted his work centered on providing food for hungry Americans, his contribution ultimately registered strongest in the arena of play. Sportsmen and women initially didn't like brown trout—they were too hard to catch—but as fly-fishing developed into more of an art than a way to feed the family, the brown trout's wily nature became prized. They represented a challenge, and they and the sport of fly-fishing in the United States grew in maturity together.

Trout offer endless seductions. The passions they inspire edge on ecstatic. Read any of the otherwise sane writers who would rather have a fly rod in hand but, lacking that, are trying to use a pencil to get the fish to rise, as if with enough rhythm, timing, and technique, the trout might break the surface of the page, shaking off commas and semicolons like so much spray, and arrive, flopping and gasping, into their laps. Maybe it's the intimacy of the connection between angler and fish, the pull of the line, muscle contracting, muscle releasing. The escape plans, the strategies, the panic of the animal are tactile as they pass through the medium of the rod. Then the fish arrives. The connection grows, palm to fin.

Many of these delights originate with the brown trout. Their natural range extended from Iceland to the Mediterranean Sea, from Lebanon to the Tigris and Euphrates. Macedonians concealing a hook in a blob of red wool that looked like a drowned insect pulled brown trout out of the rivers. So did the anglers of medieval Germany and those of the 1600s who tempted the fish with constructions of silk and parakeet feathers. Under fly-fishing pressure for longer than New World species, brown trout have been molded by natural selection into challenging prey. As they tested their mettle against humans tying flies, the ones that took the bait disappeared from the gene pool. Every innovation left fish that were warier, that fought harder. In response, the anglers perfected their art, tying more sophisticated flies, adjusting the flick of the wrist. Something about the brown trout, its stubbornness, its perseverance, its attitude, makes it seem very American to some. I recently heard a fisherman and conservationist express it this way, after describing with relish the croaking sound the brown trout makes: "Well, they aren't native, but they should have been."

꙰   ꙰

AS it now stands, most of the streams and lakes of the United States represent abandoned experiments, the laboratory after the scientist has moved on, where the equipment has gathered grime, and the reports have turned to dust. The flurry of excitement in the 1870s has taken a toll. The subjects of the experiments, the fish, work to carve out niches for themselves in the remains. Brook trout cluster near the headwaters where water is cleaner and colder; brown trout, more tolerant of warm temperatures, gather farther downstream. In other areas the fish jostle for resources. Brook trout, native to the eastern part of the United States, outcompete cutthroat trout in the West. Western stream lovers curse them. Rainbow trout, native to the West, are shoving brook trout out of the way in the East. Eastern stream lovers hate them. In some cases, they've hybridized, brown trout varieties from small German streams with brown trout varieties from Scottish lakes, cutthroat trout with rainbow trout, brown trout with brookies. Tracking native strains becomes more and more difficult. Both state agencies and individuals, dubbed bucket biologists, keep stocking the waters with exotic fish. In hot spring-fed pools in southern Montana, tropical fish dart behind rocks as ice coats blades of grass along the banks. Locals blame it on college students. Tricksters keep introducing piranhas, hoping for blood, but the piranhas die before they can cause too much trouble.

In some instances the competition between native and nonnative species is clear. In Yellowstone Lake introduced lake trout eat the native west slope cutthroat. Interactions don't get simpler than that. But in the case of other possibly competing species the picture is more murky. In one experiment on the Au Sable River in Michigan a scientist donned a wet suit and snorkel and crawled along the river bottom for two to three hours at a stretch, recording fish positions. He didn't see brown trout grabbing more food than brook trout, but he did notice that the brown trout seemed to claim the best resting places, protected patches of slow water where the fish don't have to struggle against the current to stay in place. The impact is hard to gauge.

Tracing the records of nativity is even more challenging. What were lakes like before the managers improved them? At Bootjack Lake, near Glacier National Park in northwestern Montana, the Montana Department of Fish, Wildlife and Parks recently dripped rotenone into the lake, killing everything that breathed with gills, cleaning it out to start fresh. FWP took this drastic measure because someone had planted pumpkinseed—a small orange fish with neon blue markings—in the lake, and they were outcompeting fish like rainbow trout. FWP wanted to empty the lake and restock it. But the rainbow trout weren't native either, just more popular with anglers than the runty pumpkinseed. "What fish were here before stocking started?" I asked one of the scientists. Originally, he said, noting the insufficient gravelly banks, poor sites for spawning, there probably weren't any fish at all.

Perhaps rather than a laboratory, the pools present a miniature fishery exhibition of their own. This fish is from the West Coast. That carp, stirring up the mud, started out in Germany. It chases a Brazilian snail dragging itself along the bottom, up and over Eurasian milfoil. The lacewing preparing to land on the bank evolved in Australia. The grass that bends under its weight feeds cattle in Yorkshire. They dart and dive around the water, their own Crystal Palace, reflecting their wandering histories.

CHAPTER NINE

THE BUG HUNTERS

ONE can't blame Benjamin Walsh for being a bit cranky. As the head entomologist for the state of Illinois Walsh knew all about the exotic bugs that were eating their way through American crops of the 1860s. The currantworm, the Mexican bean weevil, the citrus mealybug, the European red mite, the oystershell scale, the elm leaf beetle, and the Hessian fly pummeled the agricultural economy, bankrupting some farmers and turning others to desperate measures. In attempts to thwart the insects, they built nine-foot-high walls around their crops, sank money into potions promising miracles, sprinkled urine on their grain, and tried other questionable remedies. One entomologist, after hearing a recommendation that orchard owners drill holes in their infested trees and plug them with sulfur, raged: "Methought he ought to have added that the hole should be made with a silver bullet."

Walsh, scientific by profession, peppery by nature, had a similar lack of patience for these snake oil solutions. He was beginning to think that the cure for the plague of exotic bugs might be a dose of foreign medicine. The pests flourished in the United States, he reasoned, because they were free from the natural enemies that kept them in check at home. Without constraint of predators, they could breed themselves to pestilence in the New World while remaining only a minor nuisance in their native Europe or Asia. If explorers could seek out the insects' natural predators, capture them, and ship them to the United States, the problem might be solved. But even before he imported so much as one larva, Walsh anticipated resistance from the ruling forces of irrationality. Entomology as a respectable profession was in its infancy, and the last thing scientists needed was an idea that would make them laughingstocks. In 1866, in an article for the *Practical Entomologist,* a publication he edited with his protégé Charles Valentine Riley, Walsh wrote:

> The simplicity and comparative cheapness of the remedy, but more than anything else, the ridicule which attaches, in the popular mind, to the very names of "Bugs," and "Bug Hunters," are the principal obstacles to its adoption. Let a man profess to have discovered some new Patent Powder Pimperlimpimp, a single pinch of which being thrown into each corner of a field will kill every bug throughout its whole extent, and people will listen to him with attention and respect. But tell them of any simple common-sense plan, based upon correct scientific principles, to check and keep within reasonable bounds the insect foes of the Farmer, and they will laugh you to scorn.

The idea was not completely new. As early as 300 B.C. the Chinese collected parasitic ants and built bamboo bridges from citrus tree to citrus tree, so the ants could eat harmful insects while traveling from branch to branch. In 1762 Count de Maudave brought the mynah to Mauritius to eat red locusts. People have kept cats to eat rats since the pharaohs reigned. But in Walsh and Riley's time, the introduction of natural enemies had never been tried on a commercial scale in the United States, and its success was to take the public by surprise and turn entomologists and bug hunters into heroes.

⚹ ⚹

AS Walsh contemplated pestilence, farmers in the Far West rocked back on their porches and breathed in the fragrance of their fruit, growing sweet and heavy in the California sun. So many species flourished, fed by rains from the coast and the heat of inland valleys, that those who came West for the mining stayed for the agriculture. In contrast with the rocky fields and tired earth of New England, this rich land fairly heaved with possibility. As the mining petered out, the California story shifted: Gold no longer rolled down the rivers; it dangled, juicy, on the trees. Aspiring agriculturists knelt in the soil and tucked in seeds of grapes for wine, mulberry trees for silkworms, and poppies for opium. Citrus trees, brought to keep miners scurvy-free, unfurled their shiny leaves. Farmers planted avocados from Nicaragua, prunes from France, and melons from Turkey, then let the Mediterranean climate work its magic.

But the plants didn't travel alone. Nestled in the veins of the leaves, tucked in the curve where stem grasps branch, or crouched in a forest of anthers, stowaways lurked. Eggs, caterpillars, tiny flies—all entered the Golden State unnoticed, clinging to the plants they had evolved to eat. This rush of exotic plants (almost six hundred nonnative species and varieties of trees arrived between 1810 and 1942), planted in large fields devoted to only one crop, generated plague after plague of insects, more intense than those witnessed in the East. No one knew where the next attack would come from, and everyone jumped at an unfamiliar rustle in the leaves.

In 1868, two years after Walsh's article appeared in the *Practical Entomologist*, a sugar refiner picked a small white scale insect off his Australian acacia in Menlo Park. He captured one and sent it to a scientist in San Francisco, who passed it on to Charles Valentine Riley, now chief entomologist for the state of Missouri. While the bug men puzzled over the identity of the puffy white newcomer, the scales picked their way through an environment stocked with familiar and edible trees: eucalyptus, acacia, orange,

and lemon, all from Australia. Riding the wind, they spread to backyards in San Raphael, San Jose, and Santa Clara. A San Francisco nursery sold infested plants to a customer in Los Angeles, and soon Southern Californians were cutting back and burning their trees in an attempt to purge their gardens of the pest.

Before long, orange farmers noticed their trees weakening, growing ill, and dripping with a black ooze. Clusters of insects wrapped each branch, mouthparts like long beaks stuck through the bark, sucking the sap. The masses excreted a sugary honeydew from their rears, which trickled over the twigs and branches, growing a dark mold over time. Their red-orange bodies were dwarfed by egg sacs ballooning out from behind mature females. Larger than the bodies themselves, the sacs sported white, fluted ridges giving them the appearance of madeleines, whipped cream dollops, or puffy cushions, and observers dubbed the bugs cottony cushion scales. Each sac hatched six to eight hundred tiny orange-red nymphs, prominent against the pale balloon. Males were rare. The species was largely hermaphroditic, though the farmers didn't know it. They saw only ruin. In no time the insects covered whole orchards like an unseasonable snow.

Meanwhile, in the nine years since the scale first crossed his desk, Riley had cemented his fame. Walsh died soon after writing his article, and the younger man was on his way to becoming the preeminent entomologist in the country, including a position as chief of the Bureau of Entomology for the U.S. Department of Agriculture. First hired in 1878, Riley was gone a year later after internal squabbling. By 1881, with a new administration, he was back. Riley looked more the romantic poet than the scientist, with a sweeping wing of black hair, pale cheeks, and wide-spaced almond-shaped eyes. Born in Europe and initially trained as an artist, he had turned his colored pencils to sketching insects as a teenager. After he came to the United States and focused on entomology, vivid portraits of the country's most hated bugs emerged from his study, accurate down to the last hair of a fly's leg. He earned a reputation based on his studies of the life histories of crop pests, cultivating an intimate knowledge of what he would destroy.

Like his mentor, he had no patience for fraud. When gullible friends dragged him to the spiritualists who were all the rage, Riley delighted in exposing the mediums' tricks. His intense personality earned him both close friends and adamant enemies. Some called him brilliant, passionate, and hardworking. Others thought he was an arrogant glory seeker. One colleague summed Riley up this way: "He was the best loved, best hated, most admired and most detested man I ever knew."

Between political infighting, grasshopper outbreaks, and studies of yucca moth biology, he devoted little time to the mysterious scale. But as the Californians increased their pleas for help, Riley turned his attention to the plight of the fruit growers. In the time since he received the first report, a New Zealand entomologist identified the scale as *Icerya purchasi*, a insect from Australia. His friend Walsh's idea tickled in his mind, and he brought it up in his annual report to Congress in 1886. Why not send a special agent over to collect natural predators of the scale and bring them back to munch on California's growing population?

Actually Riley knew why not. Peeved by his frequent jaunts to Europe at government expense, Congress zeroed out the foreign travel budget for USDA officials. For the time being he was stuck with local remedies.

Riley's two special agents in California both had been edged to the country's fringe by circumstances beyond their control. Tuberculosis forced Daniel Coquillett to the warm, dry West. Albert Koebele, a German immigrant, requested a transfer from the USDA office in Washington, D.C., to escape a love affair gone bad. He asked to go far, far away, and Riley shipped him to the Pacific coast.

For the man who would be hailed as the savior of the California citrus industry, Koebele didn't look the hero's part. A small man, he lacked the radiant charisma of Riley. His clothes were rumpled. His mustache splayed in an unruly brush at both ends. In fact he looked rather like what he was, a meticulous preparer of insect specimens who kept his eyes trained on twigs. With dark hair and eyes and a short forehead, Koebele gave the impression of a small bird of prey, a kestrel that seeks out and pounces on grasshoppers in the dirt.

This very precision attracted Riley. The first time he saw Koebele's work at a meeting of the Brooklyn Entomological Society, he offered him a job. Taking Koebele as a protégé of his own, he corrected his employee's halting English, offered to let him house-sit while he was out of town, and repeatedly complimented his way with dead insects. He later wrote to Koebele, "I have not yet met anybody who pins more satisfactorily or more to suit me than yourself."

At Riley's request, Coquillett and Koebele investigated the cottony cushion scale from every angle. They observed the density of scale clusters on pomegranate, quince, and eucalyptus trees. They counted eggs in the fluted sacs and measured the joints of larval antennae. Unearthing pupae from cracks in the dirt and pulling adults from underneath bark, they picked them apart, recording every detail. Then they sprayed them with toxins, and sprayed them again, noting whether and how they died. Kerosene emulsions, whale oil soap, caustic soda, and a tobacco wash rained down on the scales, but they took refuge in leaf curls and split bark. Most of the carefully prepared concoctions merely soaked into the ground.

While his agents collected field data, Riley worked on Congress. He included a plea for a trip to Australia in each of his annual reports. The California State Board of Horticulture backed him up with a petition to the federal government outlining the state's citrus woes and suggesting Congress earmark fifty thousand dollars to introduce predators of harmful insects. Finally, in 1888, Congress agreed that Riley could send an agent to the

International Exposition at Melbourne. He was supposed to represent the Department of State at the exposition, but everyone knew his real mission would be bug hunting.

Koebele left for Australia aboard a steamer on August 25, 1888. His particular instructions were to gather *Lestrophonis icerya*, a small fly reported to breed within the bodies of the cottony cushion scale. As the larvae emerged, they fed on and ultimately destroyed their host. Before he left, Koebele and Coquillett had already received two shipments of the fly from Frazier S. Crawford, a photolithographer in the office of the surveyor general in Adelaide. Riley was particularly enthusiastic about its potential.

Once in Australia, Koebele looked up Crawford, who offered to help him with his search and give him more of the parasitic fly. Then, sometimes with company and sometimes alone, Koebele did what he had come to do: He scoured the countryside. He kept his eyes open and found something of interest everywhere. Several cottony cushion scales sucked sap on the grounds of the town hall in Sydney. At a Brisbane hotel he found a cottony cushion scale on an ornamental plant. In a graveyard tiny spiders spun webs on the scales' fluted egg sacs. While crowds elbowed through the International Exposition, admiring the aquarium and the auctioneer's block where Melbourne land had first been sold, he hunted and captured the infested scales in a nearby park. Just inside a gated courtyard, the parasitic fly swarmed on the branches of sweet pittosporum—shiny-leafed shrubs with fragrant white flowers. Koebele considered hopping the fence and filling his vials but was dissuaded by an unsympathetic policeman. As he worked, locusts flew by overhead, seeking new sources of food in an unusually dry season, 108 degrees Fahrenheit in the shade. He took it all down in a battered, leather-bound notebook, covered with fingerprints etched in dirt.

Then, on October 15, strolling past orange trees in a North Adelaide garden, Koebele saw a ladybug he didn't recognize. His companions couldn't name it either. Covered with a shiny red shell like most of the ladybugs he knew in the United States, this one had black streaks instead of the familiar spots. The designs radiated out from a line down the center of the beetle's back, like a pattern made by a child who dripped ink on a piece of paper, folded the paper in half, then opened it again to dry. Most important, as Koebele watched, a fat cottony cushion scale disappeared into the ladybug's mouth.

Near the Murray River he found more of the strange ladybugs and decided to ship them along with the flies for which Riley had such high hopes. While other entomologists could send back dead samples of their discoveries preserved on a pin, Koebele had to ship live insects in good health, in a large enough population to establish the species. To help them survive the boat ride, he created a miniature world inside a Wardian case, a weighty glass enclosure designed for growing ferns. He packed the ladybugs and flies on three live orange trees inside the case and added a healthy supply of cottony cushion scales to feed the predators and parasites on their way. He put more of the flies and beetles in various stages of development in boxes on ice so they wouldn't hatch, and together they steamed off for the United States.

Back in California in late November, Coquillett received the goods, including the 240-pound box with the trees. Ladybug larvae wandered along the outside of the glass; he guessed they had escaped through cracks in the putty. In preparation for the insects' arrival, he had built a tent around an infested orange tree. When he brought the packages inside, the ladybugs crawled to the first scale they came across. And pounced.

Chance and sloppy handling destroyed Koebele's second shipment of twelve thousand insects. Someone repacked his carefully prepared boxes and branches covered with larvae. Ice fell on some, crushing tins and their contents. Others grew mold. When Coquillett opened the dented and mashed packages on December 9, one fly, one lacewing, and one ladybug crawled out. But in

January, after Riley had put pressure on the San Francisco port to let his shipments pass through unmolested, Coquillett received additional pupae and larvae of the flies and ladybugs. These too he brought to the tented tree so they could take part in the experiment. Koebele returned with the last of the shipments, still fretting over the disastrous treatment of the squashed cargo. On the ride home he placed the insects in cold storage and asked the butcher in charge of the freezer every day how his specimens were. To the little entomologist's anxious queries, the butcher replied every day, "Your bugs are all right."

Underneath the netting, the ladybugs feasted. The bright red beetles, *Vedalia cardinalis*,* craved the scale at almost every stage of development. The larvae devoured the scales from below, while the adult beetles consumed them from above. When it came time to lay eggs, the female *Vedalia* often lifted the scales to place an egg underneath or attached an egg to the scales' puffy egg sacs, readying her offspring for their first meal. By April the orange tree in the tent was picked clean. Coquillett opened one side panel to let the ladybugs forage elsewhere, and they lifted their hard red upper wings, unfolded their large, gauzy underwings, and zoomed away.

As they spread, so did the tales of the battle royal visible to anyone with a magnifying glass. The ladybugs claimed victory in 35-acre orchards, then 150-acre orchards, then 350-acre orchards. The scales hid in plants surrounding the farms, but the *Vedalia* sought them out. Trees once covered in black slime reemerged. Plump fruit weighed down the branches. At state insectaries, people lined up with pillboxes to take the beetles home to liberate their own trees. Visiting a plot where the scale and ladybug had recently clashed, Coquillett saw only these remains of the fight: "the dry bodies of the *Icerya*, still clinging to the trees by their beaks."

---

*It was later recategorized to *Rodolia cardinalis*.

The experiment had worked. Whatever Walsh and Riley imagined when they thought of vanquishing agriculture's demons with the broadsword of science, the results exceeded their most vivid fantasy. Ladybugs were the perfect poster insects for biological control, as they shone in the popular imagination even before slaying the vile scale. Superstitions warned against killing them; nineteenth-century gardening books urged readers to treat the shiny red aphid eaters well. In England, "ladybird" was a term of endearment. Their popularity only increased in the wake of the battle with the scale. They adapted easily to the California climate and focused exclusively on the prey the USDA chose for them. The *Vedalia* made it look easy, and people wrote to Riley requesting shipments of the beetle to combat every bug that plagued them.

Just a few years before, Riley, like Walsh before him, had had to argue himself breathless to convince anyone that biological control would work, but now he needed to debunk a new myth, the ladybug fantasy. Newspapers carried drawings of ladybug species, touting them as cures for all sorts of insect ills. Some Californians with bulging wallets, imagining their entire state would soon be pest-free, stood ready to shower money on bug hunters. Meanwhile Riley tried to face down the hype. No, he said, the prey species would never be completely destroyed; no, not just any ladybug would do; no, the ladybugs wouldn't work against every pest. He even issued a warning: If the bug hunters weren't careful, they might introduce a species that would outcompete a similar native species by performing a job that was already being done well.

No one paid much attention.

With success assured, a question remained. Who deserved the credit? Mr. Craw, the horticulture quarantine officer of California, thought he merited it for first suggesting the idea of importing exotics to destroy the scale. Frank McCoppin, one of the commissioners to the Melbourne exposition, figured he had

saved the citrus industry by arranging for Koebele's Australian trip to be paid for from exposition funds. Mr. Crawford, the photolithographer who had greeted Koebele in Australia, felt his contributions as the discoverer of the parasitic fly were slighted. Riley, who now padded his annual reports with glowing letters of praise from California farmers, noted that the trip was made by one of his employees at his request. He mused about naming his new estate Vedalia but dismissed the notion as too vain. For many members of the public, the obvious answer to the question of who deserved the credit was Koebele, the brave explorer.

Honors heaped on his shoulders. Germany, his home country, referred to biological control as "the Koebele method." The California Board of Horticulture dubbed a new species of ladybug *Novius koebeli*. The California Fruit Growers gave him a gold watch at its fourteenth convention. Just in case the point eluded anyone, organization president Elwood Cooper announced, "[T]o Koebele alone is due the honor of discovery."

Koebele's career took off like a seedling in fertile ground. In 1891 he made a second trip to Australia and New Zealand, still working for the USDA but with his expenses paid by the California legislature over Riley's objection (he thought the purpose of the mission was vague). This time the entomologist traveled in style, sleeping in hotels rather than boardinghouses, never packing a lunch when he could eat in a restaurant. Reporters lined up for interviews. The articles were numerous, the laudatory statements plentiful, and Riley was rarely mentioned. Focused on ladybugs, Koebele collected one species that ate the woolly aphid, another that preyed on the red scale. Only twenty-eight beetles reached Los Angeles alive, but Coquillett did what he could with them. In later years Koebele also sent the six-spotted, Asiatic, four-spotted, and black ladybugs. Other collectors gathered ladybugs as well, including the seven-spotted ladybug, *Coccinella sempunctata*, brought over in 1900.

Koebele's fame and increasing independence grated on Riley,

his boss. Tensions still flickered between the California faction and the USDA over credit for the *Vedalia*, and Riley laughingly told a story he'd heard from Koebele about how Frank McCoppin wanted the *Vedalia* named after him. But when McCoppin caught wind of it and vigorously denied ever saying such a thing, Riley asked Koebele to confirm the story. Koebele hedged and said he didn't remember. After this the Californians slandered and sneered at Riley all the more, and Koebele didn't stick up for him, at least not to Riley's satisfaction. Jealousy intertwined with betrayal and embarrassment. Riley's letters to his man in the field grew more heated. In 1893 Koebele sent Riley a terse note of resignation and went to work as entomologist for the provisional government of the republic of Hawaii.

Like California, in its rush to replace native plants with exotic crops and garden flowers, Hawaii was crawling with nonnative insects. On Hawaii's behalf, Koebele searched through grasses and along tree limbs for parasites in Japan, China, Ceylon, and the Fiji Islands. He scoured Mexico for enemies of the lantana weed and the mainland United States for parasites of the sugarcane leafhopper and the horn fly. Like a souvenir collector, he selected a bug from every land, sure it would have its use. Meanwhile Riley, tired after struggling within the bureaucracy for so long and eager to renew his energies by getting back out in the field, quit the USDA in 1894.

A year and a half after springing himself from his job, Riley set out on a bicycle ride with his son in Washington, D.C. A recent invention, bicycles were considered newfangled, dashing, and dangerous. Of course Riley had one. Picking up speed, the entomologist hit a rock, flew over the handlebars, and cracked his skull on the pavement. Help found him bleeding from the ears, unconscious. He died before midnight.

The USDA, basking in the praise of grateful California orange growers, sent special agents in other divisions on collecting missions. A realization of the world's biological riches was taking hold, and a new breed of explorers combed the globe for living treasure. In 1898 the USDA launched a plant-collecting program by sending agent Mark Carleton to Russia. He returned with a new variety of wheat so tasty that it made the Americans wonder what they had been eating these past few centuries. In 1905 plant collector Frank Meyer took his first trip to China. Over the course of his career he brought back alfalfa sprouts for salads, lilacs for backyards, and elms for windbreaks from travels to Asia, Europe, and Siberia. He introduced twenty-five hundred new plant species in his lifetime.

But in the insect realm the more species introduced, the more miraculous the *Vedalia* seemed. Many bugs brought by Koebele and other collectors couldn't survive the climate change, traveled with predators of their own, switched their attention from the target pests to another species, or, most often, just disappeared into the waving grass. Some introductions were more successful than others. Each box swarmed with insects and hope, but none was the *Vedalia*. Koebele went to Europe in 1908 on a collecting mission and there became ill, suffering problems with the eyes that had held him in such good stead. World War I prevented his return to the United States, and even afterward he couldn't get permission to reenter the country. After paving the way for insects of every stripe and spot, Koebele could not import himself, and he died in Germany in 1924. A former mayor of Alameda, California, who tried without success to help Koebele return to the United States and failed, commented, "I feel ashamed to eat another orange."

Though the careers of both Riley and Koebele faltered after the introduction of the *Vedalia*, their legacy was assured. California bought into biological control wholeheartedly, and the tale of the rescue of the citrus industry appeared as round and bright and sweet as the fruit the ladybugs had saved.

৵ ৵

BUT nothing is as simple as it seems. The story has a sequel, darker than the original. The first problem was that, as the *Vedalia* miracle failed to repeat, people looked for other solutions. As Walsh had predicted, when the Patent Powder Pimperlimpimp made its appearance, everyone rushed to buy it. Concoctions that before the turn of the century left trees damp but still pest-filled were now being brewed to a lethal efficiency. Insecticides were coming into their own, and even California, so enamored of biological control for years after the *Vedalia* introduction, came around. One of these new chemicals had been available since the turn of the century but was promoted as an insecticide only during World War II. It was powerful and cheap, and just a pinch could do wonders. Soldiers dusted themselves with it to get rid of lice. The government sprayed it on the walls of shacks in the South to get rid of malaria-bearing mosquitoes. Foresters watched it wipe out gypsy moth caterpillars and other forest pests. It seemed as potent an evil-fighting weapon as the Spitfire. It was called dichloro-diphenyl-trichloro-ethane, or DDT.

Even before Rachel Carson's *Silent Spring* pointed out the dangers of pesticides, citrus growers noticed the first heavy outbreaks of cottony cushion scale in years after sprayings of DDT. Trees that had been clean for years dripped with black scum. Unfortunately for the ladybugs and their champions, the insecticide had a minimal effect on the cottony cushion scale but could kill a *Vedalia* two months after it had been applied. Even after DDT was banned, the dominance of insecticides over biological control didn't falter. Today in California the *Vedalia* still eat scales off orange tree branches, but the farmers lean heavily on insecticides. The balance has tipped from biological control to chemicals, and no one has much patience for the Koebele method anymore.

The second problem was that the craze for ladybugs, although abated somewhat, continued. When the Russian wheat aphid appeared on crops, the search for ladybugs to eat them was on. One of the contenders, originating in the Palaearctic, *Coccinella sempunctata* (or the seven-spotted ladybug), finally became established after repeated introductions, beginning in 1900 and continuing through the 1990s. It ate aphids with relish and spread from field to field with the help of humans who saw that colonies were placed in every state. But one study in South Dakota showed that this burgeoning population crowded native ladybug species that relied on the same limited number of aphids and pushed them into decline. Since the total number of ladybugs in the plots of corn and alfalfa remained roughly the same as the introduced ladybugs increased and the native species decreased, the scientists working on the study wondered if any more aphids were being eaten than before. An Asian ladybug, *Harmonia axridis*, first appeared in the South and then swarmed through the country by millions, gathering in masses on light-colored buildings that resemble cliffs near its native home. Big and orange rather than red, *Harmonia* edged out native ladybugs as well and spread with the help of those who hoped it would eat their aphids.

Despite all these half successes and partial failures, some biologists still carried around the well-worn *Vedalia* story like good-luck stones in their pockets. Even if they hadn't been there when Coquillett lifted the door of the tent over the orange tree, they envisioned the ladybugs bursting from the netting and heading for the infested orchards. They relived the moment that the scales and the black fungus that accompanied them disappeared like dew in the sun. To the scientists it seemed such a graceful solution. It seemed like a vindication, an affirmation, a triumph of logic and biology over the brute methods of poison and spray. To some modern entomologists, pacing weed patches in Europe and Asia looking for a six-legged answer, it still does.

CHAPTER TEN

WORDS ON THE WING

I IMAGINE him a quiet, unassuming man. While his relatives were making headlines all through the 1800s—stewing in jail on charges of bigamy, leading expeditions to the West, being captured by the Crow Indians and released moments before death, amassing large fortunes in business and giving interviews in the *New York Times* about the servant problem—Eugene Schieffelin was working for the family drug-manufacturing company, attending meetings of the New York Zoological Society, and reading Shakespeare.

But he was to have a more lasting effect on this country than any of his sisters, brothers, or nephews. They might have improved a neighborhood or fashioned a law, but Eugene changed the American landscape from coast to coast.

What was he thinking that day as he went to Central Park with eighty newly imported European starlings in cages? He must have pulled his wool coat tighter around him to protect against the cold and wrapped his scarf around his face an extra time. It was March 6, 1890, and the temperature lingered at twenty-three degrees. Snow had fallen all morning, occasionally turning to sleet and then easing back to fluffy white flakes. It made icy statues of trees and bushes in the park, softening the accusing fingers of bare twigs into gestures of pale grace. Upstate, currant and strawberry farmers worried for their crops in the unseasonable weather; in the city families hitched their horses to sleighs and prepared to go joyriding through the streets of New York.

It was a heady time, full of electricity and excitement. Sideshows and fortune tellers lured crowds to Coney Island. Chicago scrambled to raise money to hold a world's fair, and New York hoped it would fail. A businessman was arrested when his product hop soda turned out to be beer. A businesswoman launched an international matchmaking service to link the American desire for noble titles to the European desire for American money. As the world was growing faster and dirtier, hurtling with increasing speed toward a goal that was never clear, people wondered if art could save them. Poets were so revered that John Whittier instructed his barber to burn all his hair clippings to keep them from overeager fans. Robert Browning's death made front-page headlines for nearly a week. There was a movement afoot to keep the Metropolitan Museum of Art open on Sundays to draw people out of the saloons. Schieffelin, in his own attempt to civilize the beast his country had become, wanted to introduce to Central Park all the birds mentioned in the plays of Shakespeare, to offer scraps of poetry on the wing.

At sixty-four he must have had to watch his footing. Beneath the layer of snow the cobblestones of the streets were buried under four inches of mud, frozen now and slick. The birds would have been unwieldy and loud, screeching to one another in combat and fear as they pushed for space in the crowded cages. I imagine Schieffelin, carrying one cage in each hand, each bout of squawking threatening to knock him off-balance. His servants followed with the rest. Finally, under a tree that looked as if it might be hospitable when the ice melted off its branches and was replaced by leaves and buds, Schieffelin stopped and set the cages on the ground. He paused for a moment, breathing in the chill air. Then one by one, he opened the latches, and the birds stepped out into the snow-covered grass.

Dazed from months of traveling, first on rocking ships, then in bumping carriages, the starlings would have lingered near the cages at first. Some flexed their wings, still in winter plumage, flashing hundreds of white spots on black feathers. But they didn't go anywhere, just wandered a few feet in one direction, then another. At four-thirty the clouds pulled away, leaving a clear sky darkening into the deep blue of evening. Shouts and laughter filled the streets as the sleighs flew by, some carrying couples, flushed and amorous, others whisking whole families, squirming with glee. Hats and mittens littered the sidewalk. Schieffelin, growing cold and thinking about dinner, finally rushed at his birds, waving his arms and half yelling, half cheering them on: "Go, go, go." Now first one and then the whole group took off, circling higher and higher into the black sky on blacker wings.

At 2:00 A.M., just as the last sleigh bells fell silent, a pair of starlings found their way to the roof of the American Museum of Natural History and ducked out of the cold. They fluffed their feathers, preened briefly, and settled in for the night. Soon, if the hole was large enough, protected enough, they would begin to build a nest. It was almost spring after all.

Schieffelin, back in his Madison Avenue home, was most likely in bed. His wet coat and shoes crackled and hissed and sent up plumes of steam as they dried by the fire. The empty cages were stacked nearby, a few feathers still tangled in the wire mesh. Lost in a deep sleep, with the blanket pulled up to his chin, the drug manufacturer snored and mumbled and dreamed of waking to a world echoing with the same birdsong that Shakespeare had heard.

⁂   ⁂

CONSERVATION biologists now view Schieffelin as an eccentric at best, a lunatic at worst. But he was not alone in his affection for the birds of the poets or his desire to see them in the New World. Schieffelin had joined the American Acclimatization Society, a New York group incorporated in 1871. The goal of the society, laid out in its charter and bylaws, was "the introduction and acclimatization of such foreign varieties of the animal and vegetable kingdom as may be useful or interesting." Other wealthy businessman members included Alfred Edwards, a silk merchant who used part of his fortune to build bird boxes throughout New York for exotic house sparrows. Poet William Cullen Bryant was entranced by the work of Schieffelin and his friends. In 1858 he spent an evening with Schieffelin, who had just received and released a shipment of house sparrows.[*] The small brown newcomers, hopping through Schieffelin's garden and dust bathing in open patches of dirt, inspired the poet to write "The Old-World Sparrow." The celebratory ode begins:

> We hear the note of a stranger bird,
> That ne'er till now in our land was heard:
> A winged settler has taken his place
> With Teutons and men of the Celtic race.
> He has followed their path to our hemisphere—
> The Old-World sparrow at last is here.

Bryant went on to outline the fear budding in the hearts of the armyworm and the Hessian fly now that the sparrow would peck at them mercilessly. He envisioned bountiful fields of pest-free apricots and nectarines with jaunty English birds hopping through them.

Throughout the United States, from Cincinnati, Ohio, to Portland, Oregon, other acclimatization societies imported and released birds they thought would benefit or improve the landscape. Before Schieffelin, bird lovers freed starlings in New Jersey

---

[*]The source for this anecdote (*Forest and Stream*, vol. 17, no. 3 [1881], p. 43) records that Bryant was visiting W. H. Schieffelin, a relative of Eugene's. But since Eugene is often given credit for introducing English sparrows as well as starlings (many people released sparrows, so it's difficult to assign absolute responsibility), and Eugene lived on Madison Avenue, where Bryant's visit took place, it seems likely that Bryant was visiting Eugene, not W. H. If, on the other hand, Bryant spent the evening visiting W. H., maybe introducing exotic species was a family pastime.

in 1844 and in Oregon in 1889, but the birds hadn't thrived. House sparrows were introduced at least twenty times between 1850 and 1900, including a release of one thousand in Philadelphia by city officials. In 1881 the popular turn-of-the-century naturalist John Burroughs received a shipment of skylarks from a friend in England and wrote back: "Only seven out of the 24 sent reached me, and two of those died on my hands. The rest I let out on a field back of the hill, and two of them, at least, are still there, and, I think, will breed. When you come over I think you can hear the original of Shelley's skylark."

In 1871 James Edmund Harting wrote *The Ornithology of Shakespeare*, which lists all the birds that appear in the plays and sonnets, as well as the quotations that name them. While sparrows, larks, and nightingales twitter their way through play after play, the only time the starling is named is in *Henry IV, Part One.*

In the crucial scene King Henry demands that Hotspur, a passionate and willful soldier, release his prisoners, but Hotspur refuses. The enemy has captured Hotspur's brother-in-law Mortimer, and Hotspur is withholding his prisoners until the king agrees to pay the ransom. The king loses his temper, declares Mortimer a traitor, and instructs Hotspur never to speak of his captured brother-in-law again. After the king leaves, Hotspur fumes:

> He said he would not ransom Mortimer,
> Forbade my tongue to speak of Mortimer,
> But I will find him when he lies asleep,
> And in his ear I'll hollow "Mortimer!"
> Nay,
> I'll have a starling shall be taught to speak
> Nothing but "Mortimer," and give it him
> To keep his anger still in motion.

Maybe Schieffelin should have read his beloved bard more closely. In Shakespeare's presentation, the starling was not a gift to inspire romance or lyric poetry. It was a bird to prod anger, to pick at a scab, to serve as a reminder of trouble. It was a curse.

⚘   ⚘

THESE are the ways farmers have sought to protect their crops from starlings:

Helium balloons

Roman candles

Rockets

Whirling shiny objects

Noisemakers shot from fifteen-millimeter flare pistols

Firecrackers blasted from twelve-gauge shotguns

Explosions of propane gas

Artificial owls

Airplanes

Distress calls broadcast on mobile sound equipment

Chemicals derived from peppers

Chemicals that cause erratic behavior

Chemicals that cause kidney failure

Chemicals that wet feathers in the winter and keep them wet until the birds freeze to death

None has had lasting success.

The medium-size black bird, with a glossy purple and green sheen and a talent for mimicry, can charm on first acquaintance. Mozart owned a pet starling, which he had found in a shop

whistling the theme from one of his concerti. He kept it for several years and wrote a poem in its honor when it died. Pliny wrote with admiration of a starling that could recite phrases in both Greek and Latin, and Samuel Pepys noted in his diary that he had witnessed "a starling which do whistle and talk the most and best that I have ever seen anything in my life." The very name starling calls to mind a creature of the night sky, of the heavens, almost divine. But as their numbers increased in America, along with an aversion to exotic species, starlings' popularity plummeted.

When Schieffelin died in 1906, starlings were nesting outside the Museum of Natural History, in a church on 122d Street and Lenox Avenue, and in a high school in Brooklyn, but they had not yet reached Kansas, California, or Alaska. From the eighty starlings that Schieffelin set free in 1890, and the forty more he added in April 1891, the number in North America has now grown to two hundred million. They decimate fruit crops and outcompete other birds that nest in holes, including eastern bluebirds, northern flickers, great crested flycatchers, and red-bellied woodpeckers. A single flock of starlings, called a murmuration, can grow up to a million or more birds, blanketing the sky with darkness and the ground with excrement. They thrive in cities, along highways, and at garbage dumps. Several years ago an Ohio town hired an extermination company to poison the birds that were roosting nearby and making the sidewalks slick with droppings.

In 1906, when starlings had just reached New Haven, Connecticut, Frank M. Chapman wrote in the American Museum of Natural History's magazine, "From the bird-lover's point of view, the Starling is a decided acquisition to the bird-life of our cities, where its long-drawn, cheery whistle is in welcome contrast to the noisy chatter of House Sparrows." Only thirty years earlier the same clamorous house sparrows had been hailed as crop saviors by both farmers and poets. Before long, though, sparrow verse shifted its tone. In sporting magazines, in which letter writers offered tips on destroying sparrows with poison and explosives, a satirist mocking Bryant contributed "The Old World Nuisance" in 1881. It went:

> The Poet may sing in the sparrow's praise,
> But our great ornithologist, Dr. Coues, says,
> In language of truth and very plain prose,
> That the sparrow's a nuisance and the sooner he goes,
> The better we're off, so to me it's quite clear,
> That the Old World sparrow is not needed here.
>
> He defiles our porches, there's no denying that;
> He has ruined my wife's dress and spoiled my best hat.
> He hangs round the bird cage to pilfer the seed,
> And gives the canary a foul insect breed.
> He never eats worms, let us tell it abroad,
> This Old World sparrow is a terrible fraud.

The author was Fred Mather, who three years later reaped glory for importing German brown trout and sprinkling them in streams all over the United States.

The starling's glow didn't last much longer than the sparrow's, however. Even by the turn of the century some doubted the wisdom of importing all these new species on the basis of whim and desire. In 1898 T. S. Palmer, assistant chief of the U.S. Biological Survey, wrote a paper detailing the dangers of introducing animals wantonly and suggesting that species introductions should be restricted by law. Two years later Congress passed the Lacey Act, which allowed the secretary of the interior to designate certain wild animals as dangerous to humans, wildlife, or agriculture and prohibit their importation. Both starlings and house sparrows made the list. Starlings introduced in Australia and New Zealand had already proved themselves voracious crop destroyers and persistent pests. By 1940 even Chapman was noting that the starling shoved out some equally attractive native birds and that

its song, when sung by a thousand-bird chorus, no longer sounded so cheery.

If starlings have a noteworthy genetically programmed personality characteristic, it is aggression. They wait until other birds have created cavities for nests, then harass the architects until they abandon the site. Sometimes a starling enters a hole while the owner is gone. When the bird returns, the starling leaps onto its back, clinging and pecking it all the way to the ground. Even when it has claimed a nesting cavity, a starling may continue to abuse other birds breeding nearby, plucking their eggs out of the nest and dropping them in the dirt. One ornithologist watched a starling dangle a piece of food in front of the nesting cavity of a downy woodpecker. When the young woodpecker reached out of the hole for the bait, the starling dispatched it with a quick jab of the beak.

This violence begets violence, and not just from farmers. Bird lovers watch in dismay and anger as the native species they cherish are chased off by the pesky intruders. In an article titled "Nest-Site Competition between European Starlings and Native Breeding Birds in Northwestern Nevada" in the scientific journal *Condor*, Norman Weitzel described a study conducted on his property just south of Reno. Two cottonwoods on his land hosted fourteen pairs of breeding native birds in 1978, but in early 1979 a starling couple moved in. Then a dozen more joined them. As Weitzel watched through binoculars from his kitchen window, the starlings scared off American kestrels, northern flickers, olive-sided flycatchers, and house wrens in March. In April and May mourning doves, tree swallows, and house finches approached the cottonwoods, only to be rebuffed. By June the trees were offering refuge to nothing but starlings. After five years of starlings' flocking to his land, Weitzel decided it was time for a little experimentation. His scientific method was simple: He took his twelve-gauge shotgun, blasted all the birds in the trees, then counted the forty-seven starlings that tumbled down. As a result of his study, seventeen pairs of native species nested in the cottonwoods in 1987.

⚬   ⚬

AT the point in history when Schieffelin hatched his plot, America's relationship with Britain was riddled with ambiguity. America, like a younger sister, admired its older sibling's poise and experience but chafed at its patronizing tone. Thoreau and Whitman valorized the species of plants and animals in the United States as wilder and heartier than their tame British counterparts, but American thoughts still dwelt on Britain's glories. As a young nation, with many people so new to the land, America had a shortage of stories that took place on its own soil. Without a literary tradition Americans didn't know what could happen in their landscape. Was there romance in America, or did lovers need to waken to the song of the lark to experience the joys of Romeo and Juliet? Could Antony have called Cleopatra "mine nightjar," or was "mine nightingale" the only appropriate endearment? The desire to see ourselves as heroes and heroines of stories that we know and love easily translates into a desire for the artifacts of those tales. The architects of Yale University paint the stone walls with acid to give them the aged look of Cambridge, England. A traveler clips a sprig of heather from the moors where Heathcliff may have roamed and plants it in her garden in Arizona. Children at Disneyland hop on Mr. Toad's Wild Ride and watch the landscape of *Wind in the Willows* career past.

The New York drug manufacturer's error was on such a grand scale because he underestimated the dark potential of both language and biology. Wanting to release all of Shakespeare's birds indiscriminately, because they were part of the landscape of poetry, he viewed them as pleasant or, more important, as benign. Birds and poetry were natural mates. They uplifted the spirit. They kept people out of saloons. Nature and art, particularly when beautiful, are often viewed as purely good, a misunderstanding of both. Left unrealized are the dangers of dreaminess, of looking at bright colors of plumage or blur of flight, rather than the ruthless engine of DNA.

In April in Central Park starlings stride through dandelions and cherry blossoms scattered on the lawns. A male puffs out his throat and disgorges a series of buzzes and squeaks that briefly drown out the children learning roller hockey in a cement basin nearby. In a mild Seattle June starlings pace front yards, probing the grass with bright yellow beaks. One ducks into a hole in the roof of an apartment building. Such a raucous cheer rises from the young inside it's easy to believe the ancestors of this family spent generations in the Bronx before winging their way west. In November, in Missoula, Montana, the imported birds huddle in a parking lot, their speckled feathers fluffed out in ragged spikes. As they settle into cities, starlings are mimicking the noises of urban life. The rumble of cars and hum of machinery may work their way into the clicks and whistles that make up the birds' repertoire. They have been reported to imitate dogs barking, doors slamming, hammers hitting wood.

Though maybe not in the way he intended, Eugene Schieffelin incorporated the birds of Shakespeare successfully into the story of America, just as surely as Shakespeare wrote the birds into the history of England when he wrote *Henry IV*. Not much is known about why Schieffelin did what he did. No journal entries. No direct quotations. Just a few facts told and retold in scientific journals, in birding guides, and in biology textbooks with the same smug sense of horror. It's a story of hubris, of ecological disaster, of good intentions gone awry. And as they pick up the cadences of urban life, the jackhammers and the squeal of brakes, the starlings in Florida, Ohio, and New Mexico are gathering the threads of the narrative. They report what America has become since they flew free more than one hundred years ago in Central Park. They are telling it to us over and over.

CHAPTER ELEVEN

MISSION TO THE NORTH

SITTING at their desks in Washington, D.C., or in their New England homes where dim light shone on polished wood, the missionaries, teachers, and government officials who had visited Alaska for a week or a year must have been haunted by images of the Far North. As grandfather clocks and fat pocket watches ticked off the sleepy minutes, they recalled the timeless glow of the midnight sun at Point Barrow. They remembered wrecked whalers dragging supplies onto the beach. Crewmen pulling powder on deck to blast a hole through the ice. The barking cry of seals as they slid off the rocks and dived into the icy water. The sharp green spikes of the aurora borealis piercing the night sky. Against these landscapes, painted in memory's vivid colors, their present surroundings may have seemed only faint sketches. Dust settled on the arms of chairs and in the nooks of couches. China balanced on bright white tablecloths. Lace curtains scraped against the panes. And the returned travelers were restless.

Many of these same images and more were recorded in the photograph album of Sheldon Jackson, a Presbyterian minister and the general agent of education in Alaska, who in his fourteen years of Alaskan adventures had seen all these and more. By 1891 two sights had come into particularly sharp focus. One was the Eskimo villages he'd passed through on his summer tours of Alaska, once full of life, now housing only a few sick and dying. The natives appeared truly on the brink, and gaunt faces and empty houses whose occupants had died over the winter made a lasting impression. As white hunters killed off more and more whales and walrus, the native people who relied on them were left starving, Jackson reasoned. Caribou numbers dropped as well as the Eskimo acquired repeating rifles, which allowed them to hunt more efficiently. What despair and hunger Jackson didn't witness firsthand, others confirmed. One explorer reported, "In 1881, when I first visited the district of Norton and Kotzebue Sounds and the lower Yukon, deer [caribou] were plentiful. This past winter (1889) not a single animal had been seen within a radius of 200 miles." In 1890 a schoolteacher wrote to the commissioner of education, Jackson's boss, of the conditions he found in Alaska. "The orphan children in nearly all, if not all, of these settlements are in a most pitiable condition. . . . They are starved and subsist on half rotten fish and whale oil or seal oil with a little dry bread. . . . The old race of people here are dying of [*sic*] rapidly."*

The other image lingering in Jackson's mind was that of the thick clusters of reindeer, flank to nose for acres along the Siberian coast. The Siberian deer men, who owned these animals and herded them by following their long migrations, appeared well fed and content. Their substantial tents covered with hide and their warm reindeer skin clothes offered proof of their prosperity. A deer meat feast could be had by just rounding up the herd and selecting a choice buck. In the interior, countless numbers of reindeer grazed, offering a ready and replenishing supply of flesh, bone, antler, and hide. The two cultures were only a few hours apart by ship.

Jackson had an idea.

---

*In her book *The Eskimos of the Bering Strait, 1650–1898*, Dorothy Jean Ray suggests that the Eskimo may not have been starving after all. Unused to Eskimo customs and the seasonal nature of their food sources, the missionaries and teachers may have interpreted business as usual as a culture in danger.

✴   ✴

REINDEER, famous for pulling Santa's sleigh, and caribou, not famous for anything but generally acknowledged to be ungulates of the Far North, are the same species (*Rangifer tarandus*) but distinct varieties. While they are physiologically similar, their histories with humans are very different. The term "reindeer" refers to the varieties living in northern Asia and Europe that were domesticated for hundreds of years, while "caribou" indicates the animals of Alaska and Canada that never were tamed. The reindeer from Siberia, smaller with shorter legs and more varied fur than their North American counterparts, are known as *Rangifer tarandus tarandus*; the barren ground caribou that live in Alaska are called *Rangifer tarandus granti*. So Jackson's plan involved not so much bringing in a new species as replacing a wild variety with a domesticated one.

Both reindeer and caribou awe observers by their abundance. Accounts rarely speak of one reindeer and its individual expressions and qualities, but rather what impresses are the huge herds, thousands of animals migrating in a surge of motion that can shake the ground for hours, days. Their hooves, soft and tender for tundra walking in the summer, hard and strong for cutting into the snow in winter, leave round prints like a set of parentheses. As they move, tendons in their feet slide over bone, creating a click. Mothers and young may snort to each other, adding noise to the river of backs. When the reindeer and caribou are examined individually, their power is diluted and a certain goofiness comes to the fore. The nose is mid size, somewhere between the delicate proboscis of a mule deer and the honker of a moose. Perhaps the most impressive feature is the antlers, present on males and females alike. A large rack sweeps backward, full like sails in the wind or the hull of a ship. Of the tines that extend forward, over the eyes, one is often more pronounced than the other, form-

© Eveny 2000

ing a large, solid palm of bone that looks rather like a shovel. Their fur ranges from mouse gray to deep brown to white closer to the Arctic. Hidden within their skins most likely are the larvae of warble flies, whose parasitic fervor may in part spur the caribou's migratory impulses.

Uniquely equipped to survive winters at the top of the world, from nasal passages that prevent breath from freezing on their faces to hollow hairs that keep the animals buoyant as they plunge across rivers, these large mammals roaming in the Far North made it possible for humans to live there as well. For Eskimo of the interior, the caribou were everything. They provided meat, clothing, shelter, and tools in an area with few other resources, making settlements feasible even in the bone-chilling regions near the Arctic Circle. The tribes that lived by the coast and fed themselves with salmon, whales, and walrus anticipated the summer appearance of the caribou as the animals sought relief from biting insects in the cool breezes that came off the water. The lives of the hunters and their prey were knitted together, but the pattern was threatening to unravel.

꘏   ꘏

IF energy courses through our bodies like an underground stream, Sheldon Jackson possessed deep wells that never ran dry. Forging off northward every summer, dragging large mammals across the Bering Sea, sending Eskimo children to Pennsylvania, shipping Finnish families to Alaska—all these would have exhausted a small government, but Jackson thrived. Educating and converting a region still largely unmapped posed just enough of a challenge. Houses and school buildings sprang up in his wake, as if nails had broken board and rocks nestled into neat walls by the sheer force of his will. In photograph after photograph, he looks off the frame, far to the upper left, his whole head turned away, a portrait of a man propelled from the inside, following his own personal compass. Only five feet tall, he made up in intensity for what he lacked in stature. Words like "zeal" come to mind.

Born in 1834 in Minaville, New York, educated at Union College and the Princeton Theological Seminary, from the time he graduated Jackson itched to move away from the tiny leafy town and the civilized cities of his upbringing. His first impulse was to preach in China or South America, but when a doctor implied to the missionary board that young Jackson's health wouldn't stand such journeys, he tethered his ambition and set out for the western territories, seeking heathens closer to home. Soon he had established Presbyterian churches in Colorado Springs, Colorado, Laramie, Wyoming, and Helena, Montana, ministering to many small pockets of miners along the way. Then he turned his gaze south and preached in Las Cruces, New Mexico, and Tucson, Arizona. Finally, in 1877, he extended his reach even further, gaining permission from the church to civilize and educate Alaska.

During the summers he traveled from the southeastern island chains all the way to Point Barrow, setting up churches and mission schools wherever he could. By 1885 Jackson was such an expert on the area and its inhabitants that the U.S. government appointed him general agent for education in Alaska, giving him both religious and secular missions in the newly acquired territory. Every fall he came back bursting with stories of rivers, bears, and the strange customs of the natives. Conservationist John Muir was partially inspired to see Alaska by hearing Jackson speak at a Sunday school convention in Yosemite Valley. After spending a decade roaming the Sierra Nevada, Muir, accompanied by missionaries, Jackson included, hopped on a ship heading north. Muir's opinion of Jackson may be gleaned from his grumblings about the stingy missionaries and his dismay as they collected artifacts from a Native American graveyard. He writes of ministering to the ministers, answering all their questions about geology, and "preaching the glacial gospel in a rambling way."

As they took side trips to explore rivers and glaciers, one small canoe hardly seems strong enough to hold two such determined wills, two such romantic notions tugging in opposite directions. One kept his beard religiously neat; the other let his flow unruly down his chest. One hoped to bring salvation to the wilderness; the other sought to find it there. One envisioned reindeer pulling sleds and carrying mail to Alaskan villages; the other one pictured wild caribou growing fat on the plentiful grass.

Over the course of his Alaskan journeys Jackson traveled with Michael Healy, captain of the *Bear*, part of a fleet called the revenue cutters, precursor to the coast guard. In befriending Healy, Jackson formed an important alliance that likely provided both the idea and the means for the reindeer introduction. Born in the South to a black mother and a white father, Healy had been sent north at a young age by his parents to avoid slavery and had risen to become one of the most powerful men in Alaskan waters. As a revenue cutter captain, Healy, also known as Hell Roarin' Mike for his quick temper, was one of the few representatives of U.S. law in Alaska. He chased down seal poachers and whiskey smugglers (it was illegal to sell alcohol to the Eskimo), arrested murderers, and conducted the census. He also was well connected. Before taking command of the *Bear*, he had worked on the U.S. revenue marine steamer *Corwin* with Charles H. Townsend, a zoologist who wondered as early as 1885 about the feasibility of importing tame reindeer from Siberia to Alaska.

The notion appealed to Jackson for a number of reasons. Fresh from the West, he had observed how life unfolded out there as civilization leaked into the emptiness. When Jackson first took his Bible and traveled from small town to mining camp, the stagecoach passed hundreds of thousands of buffalo. By 1891, as Jackson planned to buy his first reindeer in Siberia, the remaining buffalo existed only in isolated patches of fifty or so. Vast reaches of prairie stood ungrazed, and the plains tribes tried to feed themselves by gathering and selling bones. With the buffalo gone, whites found the hungry Native Americans a heavy responsibility, and they searched for ways to avoid similar problems in Alaska. The *New York Times*, catching wind of the reindeer plan, editorialized, "[I]t has been seriously represented that unless something is done for a new source of supply the natives may have to be supported by the United States government like the Indians of the plains when their buffalo was exterminated."

On a more positive note, the missionary had seen what he viewed as a waste of plains in North Dakota and Montana converted to productivity, at least for white ranchers, by cattle. In his view, stocking the tundra with reindeer would create a similar transformation. Useless would become useful. His efforts would cause "those vast, dreary, desolate, frozen, and storm-swept regions to minister to the wealth, happiness, comfort, and well being of men." Domesticated animals—easy to care for, under control—would replace wild ones—unpredictable and prone to extinction.

In fact exchanging wild caribou for tame reindeer cut to the heart of Jackson's theological purpose. Townsend, one of the first to suggest bringing reindeer to Alaska, picked up their religious potential from the start. "In our management of these people, 'purchased from the Russians,' we have an opportunity to atone, in a measure, for a century of dishonorable treatment of the Indian," he wrote. Jackson put a different spin on the moral character of offering domesticated animals to the Eskimo, viewing their shift to agriculture as part of the conversion process. A hunting culture, viewed as "primitive" by Jackson and many of his contemporaries, would change into a herding culture, moving up a rung on the ladder of civilization. Like no other single action, the transportation of reindeer served religious goals for both importers and beneficiaries: charity, salvation, and atonement for the loss of the buffalo. With every hairy introduction, Jackson would pay heed to his calling.

⊁ ⊁

FIRED to go on the reindeer scheme, but initially denied funding from Congress, Jackson launched a newspaper campaign in 1891. The *Mail and Express* in New York, the *Boston Transcript*, the *Philadelphia Ledger*, and several Christian newspapers carried his request for funds to buy reindeer at $10 a head. The public responded to his urgent pleas. Small checks for $5 and $10 stacked up, accompanied by letters explaining how the writers could barely spare any money, but they'd heard about the fund-raising effort in church and felt obliged to help "the poor Eskimeaux." An elementary school in Baltimore raised and sent $1. Young ladies at a seminary collected $50 and mailed it in. Another $15 came from a sympathetic official at the Crow Indian Agency in Montana. By midsummer Jackson had $2,150. Even before the final numbers were tallied, Jackson and Healy sailed to Siberia, their ship, the *Bear*, stocked with dishes, beads, traps, guns, and tobacco to trade for deer.

Even though the Siberian reindeer were domesticated, they inhabited a society far from the cattle ranches of Texas. To the minister and the ship captain, the Siberians' treatment of their herds appeared shrouded in superstition rather than sound principles of animal husbandry. When Healy wanted a deer slaughtered for meat, a Siberian deer man shooed the white men away from the family circle, then led one animal apart from the herd. The reindeer's owner faced east, began to pray, then gave a signal. The deer was stabbed in the heart with a knife and sank to the ground. When it was dead, the owner gathered hair and blood from the carcass and threw it to the east, still praying. For Jackson, Captain Healy, and his crew, the barnyards and cow pastures of the lower territories probably never seemed farther away.

Obtaining a dead reindeer was the easy part. Rumor had it that the Siberians believed selling a live deer was bad luck. Because of their superstitious nature, they would never part with one still breathing, reports said. Another, less mysterious reason for their reluctance may have been that the Siberians used the

reindeer skins to trade with Eskimo and didn't want this business
to disappear. Jackson dismissed these concerns as "selfishness" and
evidence that "they have no knowledge of such a motive as doing
good to others without pay." He would not be denied.

In spite of the predictions of naysayers, the *Bear* left Siberia
carrying sixteen reindeer and deposited them on Amaknak and
Unalaska islands to overwinter. Then Jackson headed back to
Washington to tend to other business and wait out the bad
weather. During the fall Healy wrote to tell Jackson that the small
herd, especially a favorite that had strayed and then come back to
the fold, was doing well. Her name, the ship captain decided, was
Bessie.

Back in the States Jackson campaigned for a larger govern-
ment-funded reindeer introduction on the mainland the next
year, 1892, telling more tales of the beauty and hardship of life on
the ice-locked frontier. The danger, though thrilling, was real.
Violence percolated not far from the surface in a frontier society
with few officials to enforce the laws. One missionary had already
been murdered, and another tarred and feathered for interfering
with the business of whiskey smugglers. At the end of May, as
Jackson prepared for a second reindeer introduction, still without
government support, a steamer returning from Alaska reported
that Jackson himself had been shot in a scuffle with Yukon Indians who were smuggling alcohol. But like Mark Twain, Jackson
outlived the reports of his death and continued with his plans.

Early on July 4, 1892, the *Bear* glided into Port Clarence,
tucked into a bay near the tip of the Seward Peninsula. The just-
emerging sun warmed the chill only slightly, but the day dawned
calm, the water smooth. In the morning light a crowd of teachers
and government officials and Eskimo formed a curious welcoming committee, waiting on the rocky shore for the reindeer to dis-
embark. Up the rise beyond them, a flag marked the newly built
reindeer station. Salt and anticipation tinged the air. Jackson
oversaw the operation dressed in his dark fur suit decorated with

diamond-shaped insets of white fur on the hem, modeled on those of the Eskimo. Underneath the neckline a stiff white collar peeked out.

On board fifty-three reindeer breathed visible clouds in the cold air. Not long before, they'd been smelling out water lilies and marsh marigolds and nibbling birch shoots and mushrooms in their Siberian summer pasture. Then they had been lassoed and hobbled and dropped on the beach, hauled to the water, and set afloat. As the *Bear* steamed by snow-covered rises that brought ice almost down to the water, they'd wandered around a cramped pen on board, some limping from rough treatment. Now the process ran in reverse. The crew again bound each reindeer's legs with straps, wrapped a piece of fabric around each animal's body, then attached each harness to a pulley that hoisted the reindeer up off the deck and dangled it, gawky and straining, high in the air. Once the animal touched land, men carried it on a stretcher like a wounded soldier from the beach up to the station. There they removed the straps that bound the reindeer's broad hooves. Freed, the bulls and cows charged off in all directions in a burst of release and panic, thundering half a mile or more over rocky ground before one by one returning to the familiar scent and warmth of the herd.

Somewhere farther along the coast the caribou had come from Alaska's interior to let the salt air blow away the mosquitoes, warble flies, and botflies that were fired with bloodlust this time of year. Thousands of years ago their ancestors had walked to North America. But even now, when the Bering land bridge lay deep underwater, their cousins had come to join them.

࿐ ࿐

WHILE few Americans had actually met a reindeer in the snorting, parasite-ridden flesh, most had seen pictures of the beasts invested with magical powers, pulling a sleigh heavy with gifts.

Throughout the nineteenth century, ever since Clement Clarke Moore wrote eight tiny reindeer into his popular poem "A Visit from St. Nicholas" in 1822, they pranced across Christmas cards and poked their noses into holiday advertisements. Their clatter on the roof announced Santa Claus, and they could fly, hinting at the strange things that happened near the North Pole.

The practical work of harnessing and milking and slaughtering the animals didn't dull their legendary glow. When W. T. Lopp, a teacher who later became superintendent of the Teller Reindeer Station, first glimpsed the transported herds with his wife, he commented: "It seemed as if we had suddenly stepped into the fairy land of Santa Claus, although, when seen in the distance, the deer resembled a herd of cattle quietly grazing on a gentle hill slope in the States." Miner W. Bruce, a teacher and the first superintendent of the reindeer station, took the fantasy even further. One Christmas Eve, instructing his Eskimo students in the traditions and religious significance of the holiday, he loaded up a sled with parcels of sugar cubes, dried apples, and raisins, hitched up the reindeer, and traveled from house to house through the snow. Catching sight of the reindeer as they paused before a lighted window, Bruce pondered the vision: "It occurred to me that perhaps this was the first time in the history of civilization that a live Santa Claus made his midnight visit upon an errand of mercy with a team of reindeer, and that the Eskimo were the first to actually experience what throughout Christendom is only a myth."

But even as the teachers marveled at their mythical charges, no one forgot that this fairy tale laced with mercy and charity had a commercial side, just like Christmas. As the reindeer dug craters in the snow with their front hooves to get at lichen and tried to keep botflies out of their nostrils, the air buzzed with ideas for their use. They could carry the mail between Eaton and Nome, provide transportation from rivers to mines in the interior, rescue whalers locked in ice. An ideal blend of cow, horse, and sled dog, the reindeer offered something for everyone. They promised to be so useful, in fact, that once they were across the strait, the missionaries, teachers, and government officials had a hard time parting with them.

During the first few years the reindeer stations and missions controlled most of the herds, with native herders-in-training promised several each year. Jackson brought herders from the area in northern Finland known as Lapland to train Eskimo and missionaries alike and promised a portion of the reindeer to them. He launched his dreamed-of reindeer mail service, while other government officials turned a small island into a reindeer experiment laboratory. Taking advantage of the fact that being the same species, reindeer and caribou can interbreed, the experimenters released both varieties on the island to see if they would produce "carideer." The bulked-up reindeer pleased their creators but were too scientifically valuable for the island natives to eat. When prospectors discovered gold in 1898 and miners flooded in, private ownership of reindeer became a lucrative business enterprise.

Another problem with Jackson's vision of a new Wyoming up north was that many Eskimo didn't want to be cowboys, or cowgirls, or reindeer men and women for that matter. Reindeer husbandry involved traveling with the herd, becoming essentially nomadic and swearing allegiance to the group of wandering deer over a village or an extended family. Most Eskimo preferred not to leave their ancestral lands by the shore and travel inland. They wanted to hunt and fish as they always had, and if the reindeer offered more animals to hunt, all the better. Even those who were interested in and willing to herd left the reindeer when it was time to dry salmon and participate in village ceremonies and celebrations. The Lapps were alone in the eagerness of entire families to move from place to place as the reindeer migrated through the seasons.

But even when an Eskimo did want to raise reindeer and managed to accumulate a sizable number, the herd was never really his. Charlie Antisarlook, an interpreter and one of the first native herders in the apprentice program, received a loan of 100 deer, with the understanding that he would keep the increase and return the original number of reindeer after five years. After three years, though, the government took back its deer to bring supplies to miners rumored to be running out of food on the Yukon in 1897. When it appeared that this handful would not be enough, the War Department hired Jackson to go to Lapland and ship as many reindeer as he could to keep the miners from the brink. Along with bundles of reindeer moss purchased at a special market, herders, sleds, and 538 reindeer, Jackson sailed to New York, where he put the animals on a train for Seattle. He met up with them again as they boarded a steamer for Alaska and oversaw the launch of their march into the interior. Five months after the initial distress call went out, fewer than half the original reindeer arrived at the Yukon camps, only to find the miners well fed and far out of danger. Congress had spent two hundred thousand dollars on the scheme. It was an acknowledged disaster.

The same winter, a few months after recalling its loan, the government requested that Antisarlook give them his own reindeer as well to form a rescue mission for whalers whose ships were trapped in ice near Point Barrow. The reindeer arrived too late to help the whalers, who rescued themselves in the meantime, with the help of a young naturalist named Ned McIlhenny, who was collecting Arctic birds as the whalers straggled off their icebound boats. They had made it through the winter by eating caribou— more plentiful that year than they'd been in a long time—and whitefish, provided by the Eskimo. The reindeer arrived at Point Barrow with much fanfare, but very skinny and ultimately unnecessary. Antisarlook eventually received the reindeer back, but in the hard winter while they were gone he had watched his family and village teeter on the brink of starvation.

Jackson wanted to do this generous and charitable thing but couldn't seem to carry it off with grace. He continued to write lengthy and detailed reindeer reports for the government, complete with tables, charts, and maps that showed the interior of Alaska as a vast blank with "Good pasturage for reindeer" scrawled across it. Somewhere along the line, though, his original intention slipped out of sight.

The minister eventually stumbled into the chasm between the stated goals for the reindeer and the uses to which they were being put. Peppered by allegations of Jackson's misconduct, President Theodore Roosevelt sent an investigator from the Interior Department, Frank Churchill, to Alaska to ferret out the truth. As Churchill investigated, he compiled a long list of black marks against the minister: the conflict of Jackson's working for both missions and schools; the boosterism that made him exaggerate the success of the reindeer program; the reindeer, bought with government funds, that kept ending up in private hands; the shoddy record keeping; the lack of any demonstrable benefit to the Eskimo. In his final report Churchill tried to temper his criticism, but even when he defended Jackson, he couldn't help being painfully blunt: "I can not believe Doctor Jackson to be dishonest or intentionally guilty of malfeasance, but his zeal in the work of the church, his somewhat arbitrary disposition, coupled with what is commonly known as vanity, have brought enemies here and there who have made insinuations against him that in many cases are false." Other accusations proved to be true, however, and Churchill concluded his report, "I have been informed by friends of Doctor Jackson that his health is far from good, and that possibly his judgement has in consequence been impaired." Jackson left government work in 1908 and died in May 1909. He was buried at his birthplace, Minaville, New York.

The ownership struggle continued. Even though laws seeking to keep reindeer under native control made it illegal to sell female reindeer to whites, they managed to acquire them anyway. In the 1932 general roundup on the Seward Peninsula, natives owned 88,673 of the deer, one white-run corporation owned 34,235,

Lapps owned 615, and 2,250 were owned by others. So in this year and place a single nonnative business owned more than a third of the reindeer. The natives also had a hard time making money from the deer they did possess. White herders charged a one-dol-lar-per-year herding fee and would buy the reindeer for three dol-lars a head. L. J. Palmer, who studied the feasibility of reindeer raising, explained the economics of native reindeer ownership this way: "Well, there is no profit in the sense that he is producing something for sale, but it would mean a profit over the consider-ation of leaving the animal to die on the range." In 1940 the U.S. government bought all reindeer owned by nonnatives in order to redistribute them, but by that time even the slight interest in making a living in the reindeer industry was waning.

Arguments between owners, the clash of hunting and herd-ing cultures, and the politics behind animal importation meant nothing to the reindeer (though in the photographs that show each one being hoisted in a pulley above the deck of the *Bear*, strapped to a sled piled with furs, or sagging under the weight of a hefty Eskimo whose boots almost touch the ground, they do look a bit beleaguered). When the wild caribou numbers plum-meted toward the end of the nineteenth century, they left lichens and new willow shoots ungrazed. The reindeer moved into these areas and ate themselves into abundance. Even after Russia stopped selling reindeer to the United States in 1902, the more than 1,000 that had made it across the strait continued to multi-ply. Some of the reindeer strayed and joined the caribou in the interior, breeding and creating genetic permutations of "reinibou" and "bouideer" and "reinicar" that the "carideer" managers could only dream of. While reindeer owners mourned the loss of their stock, caribou enthusiasts worried that the hybrids would weak-en the wild *tarandus*. At one point in the early 1930s the domes-ticated reindeer population in Alaska hit 650,000, stretching from Point Barrow on the Arctic Ocean to Unimak Island at the southern tip.

But Alaska revealed itself to be more possessive of its wild creatures than the western plains were, and despite the reindeer herds, the balance remained in favor of the unfettered caribou. Eventually the reindeer were haunted by their domestic pasts. Both reindeer and caribou are dependent on reindeer lichen for winter forage to take them through the harshest time of the year. The slow-growing lichens cannot withstand heavy grazing and require years to replenish. Not as migratory as their caribou kin and dependent on herders to urge them from one pasture to the next, the reindeer overgrazed. In addition, the reindeer relied on humans for protection from such predators as the wolf. As the herds swelled and opportunities for profit shrank, people had neither the ability nor the desire to look after all of them. By 1950 the population had dipped below twenty-five thousand statewide. Many starved and were left to decay.

The Pribilof Islands, far off the coast of Alaska in the Bering Sea, offer an even starker picture of the swell and crash of a mismanaged herd. In 1911 the *Bear* deposited twenty-five reindeer on St. Paul Island and fifteen on St. George Island to provide the natives with food. With little herding and few predators, the St. Paul Island population took off. Scientist Victor Scheffer summed up the results: "By 1938 it included more than 2,000 animals—12 years later only 8!" In his study Scheffer showed that St. Paul had been picked clean of reindeer lichen, with only small patches clinging to places reindeer preferred not to go. The lack of food and a hard winter left the island littered with reindeer bones.

More recently population pressure on the reindeer has come from another direction. As the numbers of caribou in the western Arctic herd rebound, they are competing with domesticated animals on the Seward Peninsula, making it difficult for reindeer numbers to grow to the huge herds Jackson imagined. Still, the missionary's action wasn't all hype. Though the industry might not have converted the entire territory, on the Seward Peninsula and St. Lawrence Island, Native Americans herd twenty-five thousand reindeer as part of a business worth $1.6 million a year.

⋆ ⋆

TWO final pictures stand out from the tableau of sea captains, teachers, and ministers Jackson knew. In his collection are several sets of "before" and "after" photographs, demonstrating the civilizing, domesticating influence. In one, five Eskimo children taken from their homes near the Bering Strait stand dressed in smocks of animal skin, fur hoods framing their faces. They appear dazed, hands dangling by their sides. One girl pulls her fingers up inside her sleeves, out of sight. In the next picture, after a year of school in Philadelphia, the four girls stand in black Victorian dresses that reach up to their necks and down to their wrists. Little bat wings of ruffles flare out from the top of each sleeve. Heads are exposed, hair parted in the middle. Hands clasp primly in each lap. The one boy holds a hat, and silver buttons gleam down his front.

While it's clear why Jackson chose them, and what he hoped to show, another thought emerges. Photographs, like intentions, are carefully thought out, angled to catch a flattering light, lovingly designed. They are what we hope will remain. But ultimately, as life surges by, they're no more than taxonomic specimens, locked in a drawer and dusted with arsenic, clumps of feathers, well preserved and labeled, when what you really wanted more than anything was the whole whirring flock.

III

HERE AND NOW

CHAPTER TWELVE

IMPROVING THE OLYMPICS

THE mountain goat peered down from a patch of scree below Grand Pass in the Olympic Mountains. My hiking partner, Jennie, and I hadn't been able to see it as we mounted the snowy ridge, but it had scented our salt sweat as we climbed and was waiting. Fat and sleek in its summer coat, it shone all white except for black nostrils, black mouth, black-brown eyes, and two horns, each curved like a sickle moon. A long fringe of hair lined its jaw and bunched into a tuft on its chin, giving it the face of an old man, though it had the delicate movements of a young girl. It approached cautiously, then skittered away, over and over again, as if gathering the courage to come up and take a lick.

While similar in appearance to the mountain goats that lingered along Going-to-the-Sun Highway through Glacier National Park and darted across switchbacks in the Cascades, this one had near-celebrity status. It was one of the remaining mountain goats of the Olympics, a nonnative population that trampled and grazed on threatened alpine plants. The fate of this goat and its comrades was debated weekly on the radio and in the newspapers in Seattle. Park service personnel, animal rights activists, and native plant lovers could scarcely speak in civil tones to one another. The goats had a history, but very little future.

❦   ❦

SEVENTY-ONE years before, on the morning of New Year's Day 1925, a boat laden with four mountain goats had bumped against the shore of Port Angeles in Washington State. Plucked from the Selkirks in Canada, the goats had been trapped, packed, and shipped across the Strait of Juan de Fuca into the small timber town. While revelers from the night before tried to beat the morning light to their houses, and shopkeepers swept up confetti and bottles of Prohibition ginger ale, a local ranger and the state game warden heaved the awkward crates on a truck and drove them to the shore of Lake Crescent, at the foot of the Olympic Mountains.

E. B. Webster, a newspaperman and president of a mountaineering club devoted to exploring the Olympics, followed along and jotted notes. His club had been pushing for six years for the introduction of these goats, and he couldn't wait for them to settle in. The year before, two kids slated to be brought from Colorado died before making the trip. But here were others finally, pacing, bumping around in their boxes, preparing to make their debut on the Olympic Peninsula, just in time for 1925. Maybe it was a good omen.

At the start of the new year Washington State was drunk on hope. In 1923 the state led the United States in lumber production. In 1924 it produced five billion cedar shingles. Commerce, building, and manufacturing had outstripped all expectations, and 1925 was going to shine even brighter. Essay contests offered prizes for the best reason to move to Seattle. The railroads declared January Washington Month and plotted an advertising campaign to lure immigrants to the natural beauty and easy farming of the eastern part of the state.

But somehow, economic security eluded Port Angeles. As news of the towering cedars and Douglas firs reached the ears of businessmen, three railroads competed to survey routes through the town of eight thousand, but the plans fizzled as quickly as they sparked. During World War I government workers surged into the area to build a railroad that could transport sitka spruce used to manufacture airplanes from the northern forests to Port Angeles mills. But the war ended three weeks before the railroad was complete, and economic hopes dribbled away again. While they had a certain allure, the Olympics lacked many of the animals that drew tourists and their money to the neighboring Cascades. No grizzly bear. No bighorn sheep. No red fox. No lynx. No mountain goat.

Ever the newspaperman with a healthy respect for advertising, Webster saw an opportunity to promote the beauty in his backyard. Though he had grown up in Iowa, Webster felt soon after arrival that he knew the Olympics as well as the keyboard of his typewriter. He spent his free time exploring the mountains that towered in his backyard, fishing their creeks and rivers, strolling through the hills, and shivering at the bugle calls of the Roosevelt elk. The landscape, bound by the ocean, the strait, and the sound, turned in on itself with nowhere else to go. The brooding introspection bred a maze of creeks, summits, and valleys, endlessly familiar and endlessly surprising. In one of his books he describes the anticipation of a hiker like himself who sets out into well-known territory: "He wonders too if he will again find a band of elk in this or that old-time feeding ground; if there will be a deer in a certain draw; if he will again catch sight of a bear in a berry patch or on the river bar. Every bend in the trail brings pleasant memories, even if it is only of a brood of instantly vanishing valley quail."

He not only appreciated the scenery but helped define it. On one of his jaunts on Mount Angeles he glimpsed a clump of flowers with yellow petals and yellow centers pushing through an alpine rock garden. His was the first record of the plant, related to the sunflower but looking more like a frazzled daisy. Growing only on the Olympic Peninsula, it was later named for him as its discoverer, *Senecio neowebsteri,* or Webster's ragwort. He wrote and published three books detailing the glory of the landscape, tempting others to come explore: *King of the Olympics: The Roosevelt Elk, Fishing in the Olympics,* and *The Friendly Mountain.*

Despite his confident and glowing tone, Webster was occasionally ambivalent about his publicity project, knowing, as he wrote about the quiet and solitude of the mountains, that he guaranteed their disappearance. Thoughts of trees scarred by ax marks and campgrounds layered with litter made him wince, and he was glad he wouldn't live to see the effect of popularity on the area. Already he'd watched whisper-thin elk struggling through the snow, unable to make it to their winter range as ranches fenced off the valleys.

But he had no doubts about the Klahhane Club's plan to introduce mountain goats into the Olympics. The crags and alpine shrubs of Mount Angeles echoed those described as ideal mountain goat habitat. Billies could pick their way across rock terraces that now seemed to support little more than marmots. Kids could butt heads and play in patches of glacial lilies. Cougars posed a threat, but the timber wolf was on the decline. The club lobbied to declare parts of Mount Angeles a game sanctuary to make certain that the goats would thrive. The plan had the blessing and assistance of the U.S. Forest Service and the Washington State Department of Fish, Game, and Furs. With six years of planning, Webster had thought and rethought every possibility.

The ranger, the warden, and the journalist pulled over next to the white railing separating the highway from the lakeshore. The choppy water of Lake Crescent smacked against the rocks, and Webster must have felt a flutter of anticipation as he heard shuffling and the creaking of crates. He stepped back as the ranger and game warden pulled the boxes to the ground. They opened the doors and waited. And waited. Then, in a blur of white, the animals came out charging. One headed for the ranger, who was trying to snap a picture, and another aimed its horns at a nearby road crew. A third clattered down the highway. Eventually all four clambered up into Storm King Mountain, deft hooves finding footholds in the foreign cliffs. They paused for a moment, looking back over the road, as Webster scribbled frantically to record every movement for the next day's headlines. Then they turned away. Hop. Scramble. Leap. Gone.

⋇   ⋇

WHEN glaciers muscled their way into the Puget Sound ten thousand years ago, they isolated the Olympic Peninsula from the mainland. The rocky crags at the top of the mountains poked above the ice, providing a stage for evolution to create species of plants and animals unique to their pinnacles.

A nimble-footed botanist, if she didn't mind breaking park rules, could make a bouquet of flowers that only grow in the Olympics. Flett's violet, a woodland plant staking a claim in the crags, shoves its five magenta petals and heart-shaped leaves between the rocks. Piper's bellflower cups five petals around a cluster of yellow stamens, white in the center and tipped with violet. The closed buds, deeply grooved, look like five fingers on one hand, tips touching. Olympic rockmat takes root on steep slopes. Its leafy base sends out a frail pink stem heavy with white flowers. Olympic Mountain milkvetch, Olympic Mountain synthyris, and Flett's fleabane are also peninsula exclusives, along with varieties of wandering fleabane, lance-leaved springbeauty, and the sand-dwelling wallflower.

On a larger scale too the Olympic Mountains hold tempting attractions, remnants of an icy past. The Hoh River runs through temperate rain forest, past rotting hemlocks and sun-blotting spruces. On the shoulders of Mount Olympus, buckled blue glaciers inch downhill. Up the Duckabush River, water murmuring like conversation crashes and breaks against boulders, fallen trees, and river-smoothed stones. Along the Dosewallips, green light pools on the forest floor and trees muffled in moss line the trail. In the Seven Lakes Basin, bears and their cubs ravage logs for insects and graze on wild blueberries. Entire trees are sheathed in moss, and ferns sprout from upper branches. Mist creeps into the valleys, trailing a hint of ocean.

On our sunny July day Jennie and I sat in Deer Park campground, plotting our mileage. Most of my previous ramblings had been off-season, and I was surprised by the number of car campers, backpackers, and rangers pacing the hillsides. While we were deciding what flavor of instant oatmeal to have on the first day, a white-haired ranger drove up and told us to stay on the trail if we hiked up Blue Mountain because hikers were trampling the endemic plants. Then she asked us our destination and called in a reservation on her cellular phone when she discovered we didn't have one. When we got to the site at Grand Lake, large areas were roped off and marked by stakes showing hiking boots with a slash through them. The park was undergoing revegetation to encourage greenery on the lakeshore, trampled flat and bare.

⁂

THE same Ice Age forces that isolated the Olympic Peninsula's ridgetop plants formed the mountain goat into a superb climber capable of living all its life on the sheerest cliffs. Its hooves, designed for mountaineering, have rounded pads on the bottom that offer traction on the smallest crevice. The split in the center of each hoof allows the animal to grab at small rocks and ledges. Powerful shoulder muscles balanced on small feet can give the impression of a football player in toe shoes at ballet class, but they permit the mountain goat to pull itself up near-vertical pitches.

Although it eats like a goat and wears a similar expression, the species *Oreamnos americanus* is not really a goat. Related to the antelope, it probably left its ancestors behind and crossed the Bering Strait before dispersing into its natural range from Alaska through Canada and parts of Washington, Montana, and Idaho.

In hunting and travel lore, animals gain personalities that

appear over and over. The wolf is vicious and cunning, the butterfly indecisive and ditsy. But mountain goats, though they lack the character of tooth and claw, personify bravery and individuality in their choice of habitat. The image of the mountain goat valorized on the Great Northern Railway logo, captured in the Glacier National Park symbol, and repeated in literature is that of a solitary animal perched on the topmost crag.

In his 1893 essay "The White Goat and His Country" Owen Wister described a mountain goat in his sights: "He looked white and huge and strange; and somehow I had a sense of personality about him more vivid than any since I watched my first silvertip on a rotten log." In the 1925 book *Big Game Fields of America, North and South*, Daniel Singer reflected back on a goat that now graced his wall as a trophy: "I like to think of him, alone up there, with only the eagles and hawks and ptarmigan for company, and how he calmly surveyed the world below with that feeling of security and lordship." Dr. William T. Hornaday, director of the New York Zoological Gardens, described the mountain goat "as an animal to whom fear is an almost unknown sensation. He is serenely indifferent to the dangers of crag-climbing and ledge-walking, and to him, a 500-foot precipice is no more than a sidewalk to a domestic goat." Webster liked this characterization so much he included it in his 1920 book *King of the Olympics*.

In their lonely gazing over rivers and valleys and distant peaks, with nothing but fresh air and falling rock to disturb their contemplation, mountain goats seem to remind mountaineers of themselves. What rock climbers don't relish feeling at home partway up a cliff, as they leave behind the more domestic humans? But the solitary mountain kings are only the billies. The females and young goats travel separately in large groups, foraging for food, finding safety in numbers, ensuring the growth of the population. After a breeding season in November and December, a female gives birth in May to one kid or sometimes twins. After a time alone together while the youngster nurses and tries its foot on a ledge or two, and the mother becomes familiar with the kid's scent, the pairs rejoin the group. In these herds, unglamorous, unsung, and ever-expanding, lies the potential for trouble. They fight, establishing rigid, aggressively defended dominance hierarchies. They play, improving mountain-climbing skills. They also eat almost everything.

⚜ ⚜

FROM the small gathering on Storm King, Webster's goats spread south and east to Klahhane Ridge, Royal Basin, Lake Constance, Sawtooth Ridge, and Anderson Pass. In following years officials released more on the peninsula. One was lassoed near Mount Baker in the Cascades. Seven others were delivered from Alaska, traded for a few of the Olympics' Roosevelt elk. The population grew, giving birth on stone ledges, growing thick winter coats and shedding them again, scrambling onto higher and higher mountains. On thin ridges the goats wallowed in the earth, ridding themselves of insects and creating bare troughs up to thirty feet wide and three feet deep. Hillsides that used to be anchored by plants were stomped into sun-baked soil. In the alpine and subalpine areas where they preferred to wander, the goats found tasty flowers and chewed them down to the earth. Plant species that evolved without the need to adapt themselves to grazing provided meal after meal. Soon the Olympics' unique plants were disappearing, first a clump here, then a patch there. One of the goats' favorites was an all-yellow alpine flower, Webster's ragwort.

While the goats grew fat on alpine plants, a political battle that would determine their future stirred outside the mountains. Even before Webster hatched his goat plans, some thought the Olympics deserved government protection. Days before leaving office, President Teddy Roosevelt made a portion of the mountains a national monument in order to protect the elk, his name-sake. Not long afterward, badgered by timber and mining interests, Woodrow Wilson halved the allotment and gave the loggers back some of the trees. Then, in 1938, several years after Webster had died, Franklin Roosevelt came to Port Angeles for a visit and tour. The local press corps, especially those sympathetic to resource extraction, observed his enthusiasm with a suspicious eye. They were right. He established Olympic National Park in 1940, though the dedication waited until after the war. Over the next several decades, piece by contested piece, the park picked up land—a stretch of rocky beach here, a wooded hillside there. Eventually it hit its present size, almost nine hundred thousand acres of protected shoreline, ridge, and forest.

The goats seemed to luxuriate in all that space. Webster's notion that Mount Angeles offered textbook mountain goat habitat was a good one, and other peaks proved equally hospitable. Early on tales trickled in about goat silhouettes spotted high over Lake Crescent and a Boy Scout claimed he had roped one with a lasso. Then sightings became common and eventually bothersome. Fifty years after the first examples of *Oreamnos americanus* stumbled out of their crates, the goat population was increasing by 20 percent a year. If few predators lurked in the Olympics in the 1920s, most of those had disappeared by the 1970s. By the mid-1980s estimates placed twelve hundred goats in the park.

Something had to be done.

As the Olympics had now joined the rank of the national parks, the burden of solving the goat problem fell to the National Park Service. The policy of the NPS declared: "Management of populations of exotic plant and animal species, up to and including eradication, will be undertaken whenever such species threaten park resources or public health and when control is prudent and feasible." The mandate resonated in the Olympics, particularly because when Congress established the park, it specified that one purpose was to "provide permanent protection for the herds of native Roosevelt elk and other wildlife indigenous to the area." Plant–munching goats from the Selkirks and the Cascades were not included in this protective embrace.

The park service concocted plan after plan to slow the tide of kids and trampling, but each idea carried a slew of difficulties. Some scientists advocated contraception. They tried tubal ligation, vasectomy, and skin implants that temporarily prevent pregnancy. But these methods involved trapping, treating, and releasing each goat—no easy task when the target population frolicked on the precipices. Managers, tired of running after the fleet climbers, fantasized about a substance that could be administered by dart from a helicopter, saving the hassle and expense of capturing individual animals, but nothing practical offered itself. Other plans noted that the simplest, most cost-effective, and most ecologically sound solution would be to shoot and kill all the goats from a helicopter.

Even this wasn't an easy answer. While the park's mandate was to eliminate exotic species, Olympic National Forest and the Washington Department of Fish and Wildlife had to provide hunting opportunities, and goats were a prize trophy. So, according to some plans, the peninsula would be purged of goats, right up to the boundary between the park and the national forest land that surrounded it.

Many people loved the goats and all their hairy charisma. While native plant societies shrank at the thought of unique flowers being chewed into extinction and demanded the goats' removal as the responsible thing to do, animal rights groups rallied in the goats' defense. They gathered evidence to show that mountain goats might actually be native to the Olympics; a turn-of-the-century article in *National Geographic* mentioned a mountain goat sighting, and native tribes in the area possessed items made of goat hair and horn. Others argued, though, that the writer had made a mistake and that the tribes had acquired the items through trade. Native or not, goat advocates declared, the goats had been in the park almost a century and shouldn't have to pay the price of the human mistake of their introduction.

At the heart of these debates were definitions of "exotic" and "native." While hardly the most disastrous introduction of non-natives, the transport of mountain goats highlighted some of the conflicts and assumptions embedded in the language. Often a species is considered native to an area if it was present before Europeans first started settling here. But is four hundred years the right number to choose? What if the goats did live on the peninsula thousands of years ago but were exterminated before white explorers arrived? Why is a species "exotic" if it was introduced by people, rather than by clinging to the coat of a white-tailed deer or hitching a ride in the intestinal tract of a sparrow? Moreover, is returning portions of the United States to their condition prior to European settlement, while the Europeans' offspring themselves continue to crawl all over those same areas, an achievable or laudable goal? If we could answer these questions, what to do with the goats might become clear.

While pro- and antigoat factions went at each other in the pages of *Conservation Biology*, the park service methodically removed more than six hundred goats from the Olympics during the 1980s. Some ended up as research specimens; others tumbled off cliffs when hit with tranquilizer darts or died tangled in nets. In some years, as many as 60 percent of the captured adult females were mothers with milk, indicating that they left young kids behind. Still others were welcomed by states that felt mountain goat–deprived. Billies and nannies from the peninsula traveled to Hells Canyon in Idaho, Hurricane Divide in Oregon, Bald Mountain in Utah, and the Ruby Mountains in Nevada. In some cases the new sites had never had mountain goats before, just like the Olympics at an earlier time. Hunters on adjacent national forest land took their share. These factors, combined with a series of harsh winters, culled the goats from the high of

1,200 to around 250 in the mid-1990s. But the park service still felt the pressure of an exotic population that could explode again.

In a shelter where Gray Wolf River, Cameron Creek, and Grand Creek came together, Jennie and I met a tired trail crew setting up for the night. Winter storms had scattered trees and branches on most trails, and the crew was clearing the path for hikers. One of the crew members, a soft-spoken man with a curly blond beard, seemed eager to chat as he prepared dinner, and we were full of questions. Over a meal of potatoes, Spam, and an egg, we told him about our goat sighting at Grand Pass, and he summed up the park service's goat policy as he understood it: "Shoot them and leave them lay."

※ ※

WEBSTER's plan worked. No need to talk about the Olympics in terms of the American Alps or to compare them with Pikes Peak in Colorado anymore; they are respected in their own right. The Elwah River and the slopes that rise above it are no longer the secret playground of a small and devoted group of mountaineers but a draw for travelers from Arizona and New York, Germany and Japan. A paved road up to Hurricane Ridge drops visitors off at the heart-stopping view of peak after peak cresting southward. On summer afternoons, trailhead parking lots fill to overflow, and backpackers wait two or three hours to catch a ferry from the mainland to the peninsula. In the winter families slide down the snow-covered slopes on inner tubes and bright plastic sleds.

If Webster could pull on his hiking boots and set off through his favorite valleys and along the ridges of Mount Angeles now, what would he see? Would he despair over the bare patches where the goats take their dirt baths? The remnants of overgrazed Flett's violet? Or would he instead be distracted by the oil tankers gliding through the strait or the clear-cuts that run straight up to the park boundary? Which would spark more regret, his desire to bring mountain goats to the Olympics or his desire to attract people?

For now the goats still wander where he wanted them, eating what they can find, challenging one another over salt licks, bedding down on southward-facing slopes. A local politician has taken their side and declared that the goat killing will stop, at least while he's in office. For now politics have preempted the effort at determining a solution based on science and history, and questions about what constitutes an exotic still hang in the air.

☙ ☙

AT Grand Pass, Jennie and I made our laborious progress up the trail, watching the next range appear over the crest. The mountain goat, whose ancestors arrived in the Olympics about the same time Jennie's and my grandparents landed in America, lingered by a pool near the peak. The icy blue water reflected a chill version of the sky. The goat grazed and stopped, then circled behind us, and grazed again. But as we topped the ridge, it grew bold and stepped right into our path. Then it turned to look at us again. For the first time, my outfit of flowered boxer shorts, black jog bra, Idaho Shakespeare Festival baseball cap, and red bandanna struck me as outlandish, out of place. Our shiny nylon packs, tipping us backward under their weight, seemed unwieldy and laughable. The goat continued to stare through oblong pupils, with dark eyes that melted to brown toward the edge of the iris, and we stared back, strangers under the sun.

CHAPTER THIRTEEN

TARZAN IN THE EMPIRE
OF SUNSHINE

THEY blamed Tarzan. And why not? The yodeling monkey boy, played by a bare-chested swimming idol, swung into audience's laps on a vine, making a chimpanzee a star and inarticulateness a prized masculine trait. During the filming of *Tarzan and His Boy* in 1939, the film crew went to Silver Springs, Florida, where the clear water of the artesian spring, held in a natural bowl 125 feet wide and 40 feet deep, made underwater filming a snap. They moved in, hired locals to move water moccasins and rattlers out of the way, and pointed the camera at the baby elephant. As it swam in the transparent water, its paddling feet came into crisp focus. Months and years later, when the rhesus monkeys began clamoring in the trees, warring across the riverbanks, baring their canine teeth at boaters, and generally creating a nuisance, whose fault could it be but those Hollywood types that used the meandering river and dense foliage of Ocala County to manufacture an African jungle?

It's definitely the better story. It's the one told to visitors to Silver Springs, as the site of Tarzan's tree house is pointed out. But stubborn facts get in the way: *Tarzan and His Boy* had no rhesus monkeys in it. Before the film crew even arrived, Silver Springs locals sighted monkeys skittering along the riverbanks, and the local newspaper reported one shot while Tarzan and Jane still lingered in a Los Angeles movie lot.

A more likely scenario was this:

Colonel Tooey, who ran the Jungle Cruise boat ride, ferrying tourists past both the dense tangle of vegetation lining the Silver River and the famous waters of the spring, needed to give his business a boost. More tourists flooded into Florida each year, and 1938 was projected to be a record season with more than three million sun seekers headed south. But the big draws were the coastal cities like Miami and St. Petersburg with their lures of shopping at boutiques, betting on dog races, taking yachts out into the bays, and rubbing shoulders with the very rich. Newspaper ads flaunting regal hotels and warm curves of beach urged visitors to "Live Longer and Better" in "The Empire of Sunshine."

While Silver Springs earned a place on the recommended inland auto tour, which included stops at the Ringling Art Museum and circus headquarters in Sarasota, Cypress Gardens at Winter Haven, and Azalea Ravine Gardens at Palatka, it remained far off the beaten path, glinting with a patina of the quaint and old-fashioned rather than the gloss of wealth and good taste. Tucked high in the ankle of the pointed foot that is Florida, Silver Springs and the rivers and forest around it were easy to overlook. Hotels stressed that their facilities were clean and fireproof, rather than luxurious or exclusive.

Adding monkeys to make the Jungle Cruise more attractive was a novel idea, but not entirely new. In those years of the late 1930s monkeys abounded in North America. In Nova Scotia 25 escaped from their crates and scampered over the docks of Halifax while their ship was waiting to bring them the rest of the way to New York. In the Big Apple a rhesus monkey, aptly named Deficit in light of the damage she would do to her owner's wallet, gnawed out of her cage in a pet shop and scampered over roofs and window ledges before being cornered in a fruit store as she pitched bananas at pursuers. Gangs of monkeys repeatedly broke free from the group of one thousand kept at Monkey Hill of the Jungleland display at the New York World's Fair, bridging the moat designed to keep them in, hopping the fence, fleeing down the subway tracks, and mining the trash cans. In New Jersey a monkey escaped from a fruit boat and, fed and plied with alcohol by sailors, lived in a junkyard near the docks. In Los Angeles when 450 monkeys ran away from the set of a Bing Crosby movie, Belmont High School students took to the streets searching for them, hoping for the promised reward of one dollar per escapee.

It's easy to see how Tooey could have let them go, some quiet summer morning when the river's stillness held an echo of winter crowds and dawn fog still clung to the water. Relatives recall that he took four rhesus pairs to an island in the Silver River, between the springs and the Oklawaha River, picturing the delight of his passengers as they spotted not just the usual fare of alligators, herons, and curlews but monkeys too.

Their fur, shades of wood, nuts, and tubers—bark gray, chestnut brown, and yam yellow—blended with the dirt, rocks, and trees. They probably vanished quickly, eager to find a breeze to wash away the stink of the transocean trip in teak cages, sometimes filthy from being reused, and the smell of musty grain, sweet potatoes, and stale bread. What a relief finally to be free again, to stretch cramped muscles, to climb trees, to hide. They were primed to play their own game of Tarzan in reverse, monkeys abandoned in the human world, forced to learn its strange ways, and carve out a home at its heart. Tooey didn't get to watch them long before the monkeys scampered off onto the island where they would wrestle, play, cavort in front of big tippers as the Jungle Cruise went by, and stay, he thought, because they couldn't swim.

Of course, they could swim and did, likely diving in as soon as he was gone, tiny air bubbles silvering their fur, pushing thin muscular arms through the river water until, dripping, they reached the banks.

❊   ❊

COLONEL TOOEY wasn't alone in thinking monkeys belonged among the twisting vines and densely woven forests of Florida's interior. Something about the layers of shadow and leafy canopy seemed incomplete without small primates swinging from branches. Those that evolution didn't provide, imagination would supply. Early travelers to the Lake Okeechobee region, south of Silver Springs and no less cryptic, came back with descriptions of a strange and beautiful place where trees drooped under the weight of exotic fruit, spiders as big as baseballs scuttled through the underbrush, and monkeys and baboons hooted in the branches. They looked at the territory through glasses colored by ideas of Africa, South America, and the South Sea Islands, dreaming up the primates and mangoes they expected to find in the jungle. In 1855 a state engineer summed up the enigma at Florida's center: "These lands are now, and will continue to be, nearly as much unknown as the interior of Africa, or the mountain sources of the Amazon."

In the 1860s, when tourism started to flourish and more travelers ventured into these unknown lands to see for themselves, the rumors proved no more substantial than spray dancing off Ponce de León's Fountain of Youth. Still, Florida revealed itself to be exotic enough on its own merits, particularly for northern urbanites, escaping the gray slush of winter in the city, whose encounters with the natural world might be a trip to the wilds of Long Island. Here, as their steamers puffed up the Oklawaha River, amateur naturalists stood on deck, dipped tiny nets into the air, and caught bizarre and shiny insects. Other passengers took advantage of the mild nights and slept outside as the boat dragged them underneath the southern stars. Alligators rested long snouts on mudbanks, looking like relics from the dinosaur age. Cypress and palms grew in almost choking closeness, bound together by creeping vines. Orange-faced parakeets chattered to one another as they darted overhead. Cranes stalked fish in the eddies, then flapped off on languid wings. The Victorian travelers in their bowler hats and flowered bonnets invariably stopped at Silver Springs and took in its main attraction, the eerily clear water.

Looking over the side, they saw trees reflected in the glassy surface, drooping palm fronds, curvaceous vines. They would also
see through the surface the strange depths where fish lurked in
the shadow of rocks and scales winked silver in the light.

Monkeys were beside the point.

But as tourism became more established and the luxurious
life at the beach resorts garnered most of the attention, other
places, seeking to compete with the muscular glamour of Miami,
began to sell beyond the state's own natural blessings. Clear
springs, plentiful sunshine, alligators, sea, and sand were not
enough, and they began to advertise a paradise that borrowed
glamour from other Edens. "The Jungle Hotel" and "Hotel
Wyoming" hinted of Amazonian adventures or Wild West ranch
life in their names. An ad for Fort Myers in the 1930s offered
"South-Sea atmosphere in Florida."

To enhance their tropical splendor, attractions displayed animals from all over the world, not caged, as in a zoo, but free-
roaming, as if they were part of the fauna of this outlandish place.
Monkey Jungle, located in Miami and launched in 1933 by a man
who aimed to have the first colony of free-ranging monkeys in the
United States, displayed gorillas, chimpanzees, orangutans, and
golden lion tamarins as well as rhesus monkeys. The visitor to
McKee's Jungle Gardens, perhaps tempted by ad copy urging "See
the most amazing spectacle in America. Weird, fantastic, beautiful," was transported to a world that had little to do with species
actually residing in the state. Pulled in a rickshaw, visitors to the
Jungle Gardens could rattle one moment by alligators, skulking
by a thicket of Florida's dense native plants, and rest the next in
the shade under oil palms from Africa, sugar palms from Java, and
cinnamon trees from Southeast Asia. Safe outside a high wire
fence, a tourist could look at hundreds of gibbons, chimpanzees,
gorillas, orangutans, spider monkeys, ringtails, and rhesus
macaques, each group living in its own loose enclosure. Other
attractions brought in monkeys and then went out of business,

leaving their imports behind. A remnant population of macaques from a theme park frolicked in a swamp outside Fort Lauderdale, and another colony farther north along the coast rose from the ashes of Tropical Wonderland, escaping across a neglected moat.

But at least in the short term, the monkeys did good work. Jungle Gardens was so realistic, in its sense of carrying visitors to some wondrous, though ecologically muddled locale, that one visitor gushed, "Tarzan couldn't tell it from the real thing."

≻ ≺

EVEN though they rejected the island, the monkeys of Silver Springs panned out for Tooey. When the Jungle Cruise captain called, "Monkey, monkey," and whistled, they rushed to the shore to catch bananas, chunks of pineapple, and monkey chow tossed from the boat. Tourists loved them. Supplementing these meals, the monkeys sought out tips of cabbage palm fronds, elderberries, and whatever else the river shore offered. Repeat visitors watched increasing numbers of monkeys darting in and out of the ash and maples, pairs grooming each other, opponents baring their long, sharp canine teeth in shows of dominance, infants clutching their mothers' fur in a relentless grip.

Over time the group split, forming two troops on the north side of the Silver River near the spring, a third troop on the south side of the Silver River, and a fourth below the junction of the Silver River and the Oklawaha River. The core of each troop was generations of related females, accompanied by unrelated males who immigrated in. Each sex had its own dominance hierarchies, parceling out privileges and deprivations passed down from mother to daughter or son. At a certain age young males left the troop and roamed in bands, establishing new territories, going off on their own, often getting into trouble. At times more than five hundred monkeys, bred from the original four pairs as well as additional releases and the occasional discarded pet, likely scampered through the area. Squirrel monkeys, set free by some unknown benefactor, added to the primates on the shore.

While the ease with which the monkeys shrugged off the outskirts of Calcutta and adopted a Florida forest may seem odd, rhesus monkeys are very adaptable. Properly known as rhesus macaques (even more properly known as *Macaca mulatta*), they live near people in their native India. Unlike gorillas, just discovered by Western civilization in the mid-1800s, the macaque genus has had a long affiliation with humans, serving as the model for the "speak no evil, hear no evil, see no evil" trinkets. As their forest habitat in India is gobbled up, they've strengthened the tie, moving into villages, lounging in temples, and slowing traffic on dirt roads. More than half the rhesus monkeys in India now live in town, and they've slid easily into other environments as well. In fact, barring humans, rhesus macaques are the most wide-ranging primate.

Disconcertingly, almost painfully, human in their expressions, rhesus macaques are difficult to describe, in the same way it's hard to capture characteristics of people in general rather than a specific facial feature or a unique gesture. Hairless faces crinkle underneath expressive brows, marked by big, bare ears. Some faces are long, drooping like eggplants below the eyes. Other monkeys have rounder snouts, making the moniker dog monkeys appropriate. At rest they sit hunched over, like old men at checkers or kids playing jacks.

On the banks of the Silver River the monkeys appear at play. Adolescents climb trees and fling themselves into the water. Females groom each other. Males hang out in groups, waiting for peanuts from passersby. Maybe people flock to see them for the same reason they would go see Tarzan: Living in the trees, unfettered by stress of the office or bills or the anxieties of home life, they are models of the human on vacation.

⋆   ⋆

THE monkeys' resemblance to humans made them more than just a tourist attraction. As they marveled at the long fingers of the chimpanzee and the tangerine fur of the orangutan, visitors to McKee's Jungle Gardens learned that the facility provided an important public service. The Columbia Medical School would be coming soon to take some of the gibbons back to New York for scientific experiments. Monkey Jungle also let biologists come and study the resident primates. Marine Studios, where manatees, sharks, and porpoises rather than monkeys attracted crowds, stressed its threefold purpose: scientific studies, public interaction, and movie making. As research subjects monkeys were in particular demand. In 1938, just as the Silver Springs monkeys came into the spotlight, scientists brought over 15,851 rhesus monkeys for experimentation. But India's threats to establish laws slowing or stopping the trade made scientists think about breeding their own rhesus monkeys instead of importing more each year.

While Colonel Tooey quietly set his handful of rhesus monkeys free, the newspapers were full of the 500 brought to Santiago Island off Puerto Rico. In a joint effort of Columbia University and the University of Puerto Rico, the 350 females, 30 males, and 90 infants were shipped across the Pacific and released on the island to breed and eat from established feeding stations until they were needed. Scientists observed the experimental colony anxiously and cheered when the first baby monkey was born on the island.

Scientists learned about infantile paralysis, tuberculosis, and malaria by studying rhesus macaques. Studies on fetal malformations produced by thalidomide and radiation sickness employed the monkeys too. The development of the polio vaccine depended on rhesus macaques, and the Rh factor is named for the monkeys in whose blood it was first discovered. In the early sixties, two very nervous monkeys shot into space on the tip of a rocket and floated among stars no human had ever seen. More recently scientists have used them in experiments seeking to find a cure for AIDS. Rhesus macaques have shown us more about the way the human body works—how it responds to toxins, disease, birth control, fear, and weightlessness, what might trigger infant abuse and cocaine addiction, and why firstborns are so stressed out—than perhaps any other animal.

Soon after the success of the Santiago Island colony, breeding facilities opened in Baltimore and at Yale. While in the labs the monkeys are usually under rigid control, at times they fall through the cracks. In the early 1970s behavioral researchers brought Japanese snow monkeys to a Texas ranch. When they wandered out through a neglected enclosure, they spread into six counties.

By 1976 Charles River Laboratories had established breeding colonies of rhesus monkeys on two Florida keys: Key Lois and Raccoon Key. These islands, more delicate than the banks of the Silver River, wilted under the pressure of increasing monkey populations. Mangrove trees, shorn of young leaves and missing strips of bark, died off. Algae flourished in the feces-rich water lapping the shore. Erosion sucked clumps of land into the Gulf and the Atlantic, taking with it habitat for roseate spoonbills and white ibis. Conservationists trembled at the thought of the monkeys' reaching other islands, on their own or by way of a hurricane.

৵ ৵

IT was a hurricane that ripped the veil from Florida's hidden exotic species, previously tucked away in backyards and caged enclosures, fenced, and out of sight, and revealed the larger problem of which free-ranging monkeys are only the smallest symptom. While Jungle Gardens and the Jungle Cruise made a living off foreign animals on display, others, like the medical facilities, kept their nonnatives hidden from view. When Hurricane Andrew stripped flimsy coverings away in 1992, peacocks strutted through empty buildings, western coral snakes slithered across the highway, and a lion reportedly roamed the streets. Hundreds of monkeys from the Mannheimer Foundation near the Everglades and the Perrine Primate Center at the University of Miami fled through destroyed enclosures. They hopped backyard fences and left homeowners rubbing their eyes for weeks afterward. It was as if the whole state were a zoo.

Like Hawaii and other places with welcoming climates and particularly unique ecosystems, Florida is overrun by exotics. Boa constrictors from the West Indies, nine-banded armadillos from Texas, the red-whiskered bulbul from Southeast Asia, sea anemones from Asia, houseflies from Africa, and horn flies from southern Europe all make their home in some Florida pool, tree, or grassy pocket. Even most of the greater flamingos, bound so tightly into the state's culture they are native in the minds of many, come primarily from introduced populations, though migrants from the Bahamas do occasionally drop by. The costs of this influx are high, with patches of uniqueness like the Everglades, tropical hardwood groves, and pine rockland shrinking like mud puddles when the weather shifts to sun.

Slowly Florida is beginning to wake into itself again. It doesn't want to be someplace else anymore, a movie set for jungle drama, a stage for a South Seas fantasy, but how to take the scraps that remain and sew itself back into a coherent shape? Those who treasure the state's natural wealth look down in dismay at the walking catfish traveling from pond to pond and the fire ants charging around pebbles and twigs. Then they look up and see monkeys.

❧   ❧

BY the mid-1980s, the tide had turned against the Silver Springs macaques, no longer viewed as an innocent diversion or a welcome addition to Florida's tropical image. They weren't causing that much trouble, nothing like the monkeys on the fragile keys, but they might, and they stood as a symbol for all the other exotic species adapting to Florida's climate. State wildlife managers envisioned monkey gangs terrorizing children or a tourist reaching out to pet one of the riverside monkeys as it pulled a banana from its skin, getting bitten, and dying from simian herpes virus.

Gradually, mostly with state sanction, Silver Springs monkeys started to disappear. Someone stole monkeys from the troop in the Ocala National Forest, but since the macaques weren't supposed to be there anyway, no one chased down the thief. Others were shipped to a zoological park in Missouri. Several hundred were trapped and sold as laboratory animals. So even though Tooey didn't intend it, Silver Springs ended up as much a breeding ground for scientific subjects as Jungle Gardens or Santiago Island.

Silver Springs remains a tourist draw. Visitors can ride on the old-fashioned glass-bottomed boats or visit a Florida of the past "untouched and untamed" on a Lost River Voyage. In addition, they can view performing parrot shows, stop by the petting zoo, and still find the Jungle Cruise, which now offers views of animals including giraffes, zebras, and gazelles on the banks of the Fort King Waterway, parallel to the Silver River. Similar opportunities to see a jumble of native and exotic species in a thirty-five-acre enclosure are afforded by the Jeep Safari.

Despite the efforts of the 1980s and several state mandates to remove, eliminate, or cage them, monkeys may still be seen skittering in the forests along the Silver River or hanging off the signs that warn boaters not to feed them. They also scamper on Raccoon Key and Key Lois, and haunt the grounds near where Tropical Wonderland used to amuse, before jungles went out of fashion.

CHAPTER FOURTEEN

SWAMP RICHES

EARLY in the spring of 1939 a large furry rodent with a hairless tail, webbed feet, orange buckteeth, and long whiskers curving down either side of her face gave birth to five young at the U.S. Fur Animal Field Station at the Blackwater National Wildlife Refuge in Maryland. Looking like a mix of beaver, otter, and rat, the mother nutria rested after her labors while Herbert Dozier, the director of the field station, watched over her and weighed the squirming pups, noting that each stared at him with open eyes. Like an anxious new mother himself, he worried over how best to care for the litter, debating back and forth with his superiors in Washington, D.C. Would the young nutrias prefer carrots, fish, or protein pellets? Was their water clean enough? Should they plant kudzu in the marsh for the grown rodents to feed on? Director of the Fur Resources Division of Wildlife Research Frank Ashbrook offered advice and urged Dozier to keep his papers a little more in order.

That same year, more than a thousand miles away in Louisiana, a gang of nutrias burrowed out of the pen where Edward Avery McIlhenny kept them and disappeared into the surrounding swamp. He had imported six pairs from South America in 1937 and brought them to his home on Avery Island in the southern part of the state. McIlhenny, known as Mr. Ned, was both a shrewd businessman and an avid conservationist. Avery Island possessed riches enough, being a natural salt mound, but to top it off, one of McIlhenny's ancestors had stumbled across *Capsicum frutescens*, a red pepper from which he created Tabasco sauce. Soon the island glowed crimson with pepper plants. McIlhenny inherited the Tabasco business as well as the salt mines and split his time among probing the island for salt, harvesting and fermenting peppers, and adding to his burgeoning collection of animals and plants.

McIlhenny was a connoisseur of the exotic. His plant collection, dubbed Jungle Gardens and covering acres of his family estate on Avery Island, included species from all over the world. Egyptian papyrus, Japanese camellias, and South Seas papayas flourished side by side. More than 150 varieties of bamboo created dense thickets, and 26 kinds of palms shaded a plot set aside to mimic the desert. In a cool glade a Buddha statue sat surrounded by shrubs and flowers from China.

As a young man itching to explore McIlhenny ventured poleward in the mid-1890s and then organized and financed his own ornithological collecting trip several years later. While in the Arctic in 1897 and 1898, he collected 1,408 bird specimens and carted hundreds of northern species to the East Coast to be analyzed. Admittedly they were dead and could do little harm.

On the same trip McIlhenny witnessed the benefits of doing business in live exotics. In 1898, the year ice trapped seven ships in the waters off northern Alaska, the twenty-four-year-old McIlhenny was at Point Barrow, working on his collections. As the whalers limped into the settlement, he turned his attention from shooting and cataloging birds to housing and feeding many of the sailors during their long wait for rescue. According to legend, the stranded men employed themselves by making blankets out of the cotton McIlhenny brought to preserve his specimens. McIlhenny and the whalers survived the winter on caribou meat, plentiful that year. Still, when Sheldon Jackson's reindeer rescue team finally galloped into town, bringing meat and supplies that were no longer needed, the young man must have glimpsed and been intrigued by the excitement surrounding these imported beasts.

Despite his attraction to nonnative species, he knew well the damage they could do. At his beloved home on the Gulf Coast, starlings stripped holly berries, hackberries, and chinaberries from the branches before the robins and cedar waxwings arrived. He shot at them and claimed they were so densely clustered he killed sixty-nine with one bullet. The raid prompted him to write an article titled "Are Starlings a Menace to the Food Supply of Our Native Birds?" in 1936. For other lessons in the dangers of importing live animals, he didn't have to look farther than his bookcase. In his extensive natural history library, along with *The Natural History of the Bible, Beautiful Ferns,* and *Narrative of a Journey to the Polar Sea in the Years 1819–20–21–22,* he could delve into the Massachusetts State Board of Agriculture's exhaustive tome *The Gypsy Moth* for a study of exotic perils.

McIlhenny also had a passion for the native species of Louisiana. In his free time he might spend all day sitting behind a burlap blind on a box, observing a female alligator on her nest and taking notes. He made his name in conservation by creating a sanctuary on Avery Island for snowy egrets called Bird City. The plumage of the birds had been so in demand for hats in the 1890s that the egrets had been hunted into near extinction. McIlhenny tracked down a few pairs and raised them into a flock of thousands. One portrait shows him with the birds that made his reputation. The egrets, perched next to him, flying behind him, and resting by the hundreds in the background, display elongated grace, thin black legs, the swoop of an S-shaped neck, and delicate curling feathers. Their thin beaks are like needles stitching a straight seam through the air. They soar and flap and fluff and preen. McIlhenny sits next to them in a heavy blazer, with a tie knotted just under the folds of his neck. Jowls round out his face, curving the lines from eyes to chin. His hair is brushed back off a domelike forehead. His whole body tilts forward, gravity-bound. Every inch a businessman, McIlhenny looks as if he might rise out of the frame, clasp your hand in a meaty paw, and offer to sell you life insurance. But he pictured himself as a snowy egret.

That he envisioned this transformation, heavy bones traded for light ones, pink flesh for white feathers, is evidenced by his book *Autobiography of an Egret*. He takes the reader through the life history of an egret living in the Bird City he created. His protagonist tells of the comfort of nestling in his mother's feathers and the anxiety of flying south for the first time. He is struck by love and lauds his mate-to-be: "Her deep lustrous eyes sparkled with the fire of youth. Her face was tinted with health, a delicate orange." McIlhenny was a keen observer, and the book, like his others, *The Alligator's Life History* and *The Wild Turkey and Its Hunting*, is accurate in its detail, if overwashed with emotion. The book has considerable appeal, though it ends on a self-congratulatory note as the egret muses about McIlhenny: "When our tribe was at its lowest ebb, he gathered together the remnants and saved us from destruction. We cannot show our gratitude, but when all is over for me and I pass on to 'bird heaven' I mean to whisper to the Keeper of the Gate, with all due respect to the One above, the name of the man who saved us." Humility was not McIlhenny's strong point.

With his Jungle Gardens and Bird City, McIlhenny transformed the family estate into a haven for animals that needed his protection, stocked with biological riches from all over the world. His salt mound became a terrestrial Noah's Ark. His motivation seemed to spring from a mixture of savior's impulse and collector's drive. Somehow nutrias fitted into this scheme. While the government scientists Dozier and Ashbrook documented their experiments with nutrias with detailed explanations in letters and notes, McIlhenny kept quiet about his plans. What exactly he wanted with the South American rodents is unclear. Whether he had profit on his mind when he imported the nutria or just wanted another addition to the menagerie remains a mystery of the bayou. Eccentric and wealthy, he didn't need to justify himself to anyone.

⚹   ⚹

WHATEVER the Tabasco King was thinking, plenty of others looked at the nutria and saw a fortune paddling through the swamp. In the wild, clumped with mud and dripping with water, nutria fur appeared as desirable as rat leather, but with the rough guard hairs plucked out and the dirt rinsed off, the underfur glowed softly and silkily. In their native lands, where they frequented rivers and coastal areas, nutrias had long been prized for their pelts, and it didn't take long for other countries to catch on. In the middle of the nineteenth century Uruguay and Argentina exported thousands of hides a year. Travelers to South America during the same time found native people wearing nutria fur and serving the meat, roasted, on a stick. As early as 1885 the National Society of Acclimatization in Paris imported nutrias, intending to breed them.

Any last reservations the stylish crowd might have had about snuggling up to these funny-looking rodents were washed away by the pure demand in the years after World War I. Furs were the rage in the 1920s, when many could afford mink stoles and raccoon coats, and lingered in popularity in the 1930s, when they symbolized the riches that for most were out of reach. Movie star glamour portraits were not complete without a full-length mink to set off a pose or at least a hood lined with fox to frame the face. The clamor for different species to turn into wraps and capes made coveting the hair of a big-toothed, rat-tailed swamp dweller seem perfectly ordinary. Much weirder things were going on.

The variety of animals used for fur at the time could fill a zoology textbook. Magazines displayed marmot jackets, beaver collars, and wolf fur wraps. Men and women draped themselves in mink, ermine, silver fox, raccoon, sable, muskrat, lynx, Persian lamb, chinchilla, ocelot, leopard, panther, monkey, skunk, kangaroo, mole, opossum, otter, gray squirrel, seal, wolverine, rabbit, bear, weasel, wombat, and goat. Other species whose skins made the catwalk had names unknown to biologists: marabou (a kind of stork), glutton (a wolverine), and breitschwantz (aborted fetuses of karakul lamb). It was as if some bone-piercing chill afflicted members of the fashionable set, and they could stop the shivering only by piling on more outlandish pelts.

But this tactic required an increasing supply of wild animals that often couldn't keep up with the demand. The search for beavers to make hats had prompted much of the explorations of North America, and as one population was depleted, trappers had always just ventured to new frontiers. But those frontiers were shrinking, and by the 1830s worried naturalists predicted the beavers' extinction. As the century unwound, more and more species kept slipping through their fingers. The Guadalupe fur seal was reported extinct five times; hunters cleaned out all but a few sea otters along the Pacific Coast. By the early 1900s, as popular concern about vanishing animals grew, trappers' and furriers' lives were further complicated by an obstacle course of restrictions and prohibitions on hunting wild animals that were in many cases already gone.

Entrepreneurs began to search for a way to supply the market without relying on prey that was growing scarcer. While fur trapping had been a mainstay in the United States for a long time, fur farming as big business emerged only at the start of the twentieth century. In the 1890s on Prince Edward Island in Canada fortune seekers captured and began breeding a red fox with a rare fur color, silver. No one paid much attention until the company released sales figures for silver fox to the public in 1910, triggering the great fox rush. A single Prince Edward Island pelt had sold for $2,627, and it wasn't long before investors paid $35,000 for a breeding pair. By 1922 the U.S. government had leased 150

Alaskan islands for fox farming.* Mink ranching didn't get under way as a sizable business until the 1920s. An article in the *Illustrated London News* under the subtitle "A Ten-Inch Rodent with a Fortune on Its Back," told the story of a Mr. Chapman who traveled to 11,300 feet in the Chilean Andes in order to trap twelve chinchillas to start his fur-ranching operation in the United States. He carried them back, panting from the heat, in a special refrigerated compartment. Chapman, grinning with a mouthful of bad teeth and looking as if he really, really needed this scheme to work out, posed for the magazine with one of the precious rodents perched on his head.

Fur farming appeared to solve many problems. No longer would trappers be dependent on fluctuations of wild animal populations. No longer would investors need to fund expensive trips to untapped frontiers. An entrepreneur needed only to import the right exotic species or domesticate a native one, build a cage, hit on suitable food, and the money would roll in. France imported silver foxes and opossums. Germany, Switzerland, and Russia sought nutrias. England brought in the muskrat. It was like a gold rush with a guaranteed lucky strike or a lottery ticket with a tipoff on the winning numbers.

In the early 1920s Fur Resources Division Director Ashbrook himself wrote an article on the growth of fur farms in the United States. Soon it seemed as if anyone who had a fenced patch of ground in his backyard wrote to Ashbrook asking about breeding mink, chinchilla, or this mysterious creature nutria. As a result, Ashbrook and Dozier's experiments in Maryland had a twofold

---

*Both red and arctic foxes were stocked on these islands with disastrous results. Many of the islands had no native land mammals, and with trees scarce, bird species nested in flocks on the grass or in burrows underground, making them easy prey for the introduced predators. Crested auklets, common eiders, storm petrels, rock ptarmigans, tufted puffins, and many other species disappeared after foxes arrived. As fur prices dropped and fox in particular was no longer so desirable, the U.S. government stepped in to try to save some of the remaining bird populations.

purpose. The first was to learn if farming nutrias for fur could be profitable. The second was to determine if setting the rodents free in the marshes would provide a windfall for trappers. Throughout 1939 and 1940 the fur operation in Maryland continued to grow. Ashbrook and Dozier bought more nutrias, guaranteed to be of the finest South American stock, from a dealer in Canada and requested several more from the Bitter Lakes Wildlife Refuge in New Mexico, where a local population of fur farm escapees flourished. Actually, nutria sightings dotted the country, indicating that they could survive just fine in North America, even without protein pellets. But Dozier continued to care for his growing pack, tenderly noting each birth, feeding the newborns from eyedroppers, assuring Ashbrook, after one had given birth to a litter of eleven, "the mother is doing nicely." Ashbrook wrote back, encouraging Dozier in his work and chastising him when the government inspectors came by the station and found beds unmade.

At the end of July 1940 Ashbrook sent a prepared nutria pelt from the Bitter Lakes Wildlife Refuge to a fur merchant for appraisal. The reply came back that the skin was worth $2.25, not a bad price, but far from the per pelt figures quoted by advertisers hawking nutrias as an investment. The handful of change was certainly not enough to sell the ranch for a pair of nutrias in the hopes of buying a mansion. The answer to Dozier's question—was breeding nutrias profitable?—would seem to be no. As Ashbrook pondered this news in D.C., maybe drafting the next discouraging series of letters to nutria enthusiasts, down in Louisiana people were keeping an eye on the weather.

⁂

THE forecasts for Sunday, August 4, called for cloudy skies and occasional thundershowers, but with a such a strange summer, who knew what to predict? July had broken records for both heat and cold, topping out in the high nineties and plunging to the low sixties. Visitors lingered at Gulf Coast resorts, watching the skies and hoping for sun. Determined weekend gardeners braved the damp to plant sweet peas and zinnias. Others holed up in movie theaters to lose themselves in *They Drive by Night* with Humphrey Bogart or *Young Tom Edison* with Mickey Rooney. Meanwhile high winds picked up off the coast of western Florida.

On Monday society matrons fluttered and clucked in a last burst of planning before the summer season wound down. The governor went to Washington to extract promises of aid for farmers whose crops had suffered rain damage. National guardsmen slogged through drills in mud and drizzle. To the east, in Mobile, Alabama, winds up to thirty-five miles per hour buffeted cars. Tides in Mississippi crept down the shore while barometers recorded unusually low pressures. And 225 miles out in the Gulf the Greek freighter *Oropos* sent out a distress call.

Overnight Louisiana learned this was more than just bad weather. A tropical storm careered along the coast, wrecking whatever it touched. Wind and water blew out piers. Waves lapped up the steps of the seawall and slipped under the doors of houses built seven feet off the ground. Gusts ripped through power lines and sent awnings crashing through shopwindows, mixing glass with merchandise. Rain streaked the new rosebud wallpaper of a dance club while a duck paddled by the piano. As the bayou flooded, trappers fled their homes, clutching a hat or a sack of rice. Muskrats, trapped by high water in their nests, had nowhere to go, while deer fled for high ground. By Wednesday morning newspapers were calling it a hurricane as ninety-five-mile-per-hour winds roared over the Louisiana-Texas border.

At the height of the storm, flooding closed highways, drowned cattle, and forced many to flee to shelters set up by the Red Cross. The southern part of Louisiana was deluged.

After the hurricane had passed, the *Oropos* made it to shore with a patched rudder, and everyone prepared to go home and start the work of cleaning up. But the rains didn't stop. For days the downpour continued, as if the clouds, fury spent, were exhaling a long, damp sigh of relief. Highways closed, living rooms flooded, and refugees trudged by the hundreds back to high ground. Red Cross volunteers gathered food and clothing. On Avery Island the rain began to fill the pen where McIlhenny kept his remaining nutrias. Dirt turned to mud, which gaped into murky pools. Drops spattered on the nutrias' eyes and snouts, running through their furs and soaking them from above, while the building flood lifted 150 of them from below. Soon their paws no longer touched the ground. They swam. The rushing water carried them, slick and buoyant, lifted them to the edge of the boards stacked one against another to make a fence, and washed them out over the land.

࿐   ࿐

WHILE many other furbearers would have drowned in the flood—dead muskrats piled high on the banks—the nutria found itself in its element. If one, maybe a female, found herself swimming in the swamp, webbed hind legs and a flat, hairless tail would let her navigate to a good spot on the shore. Awkward on land, where she scuttled around with a hunching gait, the nutria could slide into water, fresh or salt, with grace, leaving a slight, sinuous wake. Thick brown fur, coated with oil that made it shiny, kept her warm, even in ice-lined channels, but in a Louisiana August she had no such problem.

As the rain eased and the swamp drained, she discovered ideal habitat. Her food was the substance of the marsh itself. Strong front teeth, notably orange, gnawed at pickerelweed, bull tongue, arrowhead, and square-stem spike rush. After harvesting a plant, she brought it back to a chosen spot—a floating log or branch—to nibble at the roots and tender sections, abandoning the rest in a scrap heap that grew through time.

Her nest, a simple affair of saw grass or cut-grass, was probably in a cattail stand. Four and a half months after finding a male in the bayou and mating, she gave birth to a litter of fur-covered pups, ready to follow her into the stream hours later. Often they caught a ride, clinging to her shoulders or flanks, as she cut through the water like a fur-lined barge. Even while she swam, they could nurse, as her teats lined her back, several inches on either side of the spine. In four to six months these offspring would be able to breed as well.

As the weather dropped below forty degrees Fahrenheit, she prepared for winter by amending the shore to her specifications. Digging into the mud on the side of the bank, the nutria carved out a burrow ending in a chamber about five feet in and above the waterline. Lining it with vegetation to keep out the winter cold, she created a space that generations would return to, each adding to the underground network of tunnels and rooms.

While she nibbled and dug and nursed her young near Avery Island, other nutrias were doing the same throughout North America. McIlhenny's importation hadn't been the first, though offspring of his escapees probably account for most of the nutrias now living along the Gulf Coast. Encouraged by Stanford University President David Starr Jordan and the U.S. Biological Survey's C. Hart Merriam, Will Frakes had brought a male and three females to Elizabeth Lake, California, in 1899. There is no record of whether or not they increased, but others soon followed. In Canada, C. R. Partik had received some from a friend in Germany in 1928, and he toughened them by raising them in outdoor pens through Quebec winters. Ten years later the *Fur Trade Journal of Canada* published a pamphlet called *Nutria Raising*, citing Partik and encouraging others to go into the fur-farming business. They did. Accounts describe nutrias in Washington, Oregon, and Michigan in the early 1930s and in New Mexico in the mid-1930s.

Wherever else nutria talents might lie, they are master jailbreakers. All over the country stories similar to the Avery Island release surfaced. In 1941 orange-toothed swamp creatures showed up along the Sammamish River in Washington State. This strange beast puzzled locals until they made the link to nutria fur farms in Seattle, Bothell, Bremerton, and Bellingham. Other feral populations surfaced in La Conner, on the Colville Indian Reservation, and outside Portland, Oregon. In the Bitterroot Valley of Montana an irrigation ditch breached in 1944, washing away nutrias kept for breeding. Two years later trappers hauled in several and stared, mystified by their bizarre catch. Back in Maryland in 1943, nutrias escaped from Dozier's enclosures at the Blackwater National Wildlife Refuge. In following years people living near the refuge found winter-killed nutrias under their cabins, and eventually several thousand roamed wild around the experiment site. Not all the releases were the result of nutria ingenuity, however. During World War II fur's popularity disintegrated, and many former fur breeders set their unprofitable investments free. Despite these introductions, promotions, and escapes, nutrias hadn't gotten loose in such large numbers in such suitable habitat before 1940. The nutria took to the Louisiana swamp like mosquitoes to standing water.

Five years after the August hurricane, trappers took 8,786 nutrias from Louisiana marshes. The next year they pulled out 18,015. By 1961 more than one million nutrias were chewing up saw grass, flattening cattails into nests, and digging burrows into the channel banks. McIlhenny's island paradise was slipping beyond its borders.

Once the nutrias scampered through the marshes in ever more abundant numbers, no one knew quite what to make of them. True, they plowed into rice field levees to make burrows and chewed up crops of sugarcane, sweet potatoes, and cabbage. Cows tended to wander over tunnels they'd dug and fell through, while muskrats, already under pressure from developments that swallowed their salt marshes, didn't need the competition. But nutrias also gnawed at undesired plants (and desired ones, as those who tried to employ them as weed cutters soon discovered) and offered a new source of fur for trappers. In addition, their meat could be sold to feed mink, a more profitable species. Furthermore, when looked at in the right frame of mind, they were kind of cute.

This ambivalence showed up in state decrees. Shortly after their release, in 1946, Louisiana protected the newcomers as valued furbearers. But by 1958 the nutrias had eaten their way into "outlaw" status in certain areas, a position that allowed them to be killed indiscriminately. By 1963 seventeen parishes deemed the nutria an outlaw, but the next year they were protected furbearers again throughout the state. Economics undoubtedly played a role, as in 1965–66, nutria pelts from Louisiana brought more than three million dollars to trappers in the swamps. No wonder the animal had so many names. Known as nutria or coypu (when reported on in newspapers), *Myocastor coypus* (when referred to by scientists), rangodin (when served for dinner), castorino (when draped as a fur), or varmint (when nibbling at sugarcane), the rodent tugged at heartstrings, purse strings, and trigger fingers. Nutrias were introduced to new areas constantly, while at the same time hunters would build rafts, litter them with carrot chunks, shine a light out over the marsh, and open fire at the first paw that reached up after the bait, just for the heck of it.

McIlhenny died in 1949, leaving behind a beautiful garden, a city of snowy egrets, and a swamp burgeoning with nutrias. In the 1950s, even as the rodents became more and more available to anyone willing to trek through the cattails to set out a trap, nutria farming surged again. To the untrained eye, it didn't seem such a shady business. While not as popular as sable or beaver, nutria fur did show up in stores and magazines. Designers occasionally offered a sweeping nutria coat that dropped down below the knees, a nutria overcoat with pony skin collar and cuffs, or a plush brown travel coat with side slits and a tapered back of dyed beige nutria. Other breeding operations, such as mink and silver fox, had been subject to wild speculation at the beginning but had settled down into legitimate business enterprises. This was the promise dangled before prospective investors.

Organizations multiplied as did the nutrias themselves. The Superior Mutation Nutria Organized Ranchers, Purebred Nutria Associates, Inc., International Nutria Marketing Association, and Cabana Nutria Inc. (whose logo featured a big-toothed rodent wearing a crown over the words "sign of quality") all recruited new members. Slick brochures featuring models in ankle-length fur coats and white gloves assured investors one nutria pair would produce six hundred pelts thirty-nine months after purchase (small print: if each female gave birth to fifteen young a year) and offered hopefuls eighteen thousand dollars annually (small print: if each pelt sold for thirty dollars). Ads in the back-of-comic-book style hinted at annual earnings of up to fifty thousand dollars. Even if the furs didn't bring top prices now, promoters glowed, just wait until nutria color mutations hit the market. Black, white, sand, pastel, and champagne pelts would soon wash away the plain old brown, bringing with them waves of cash.

But nutria fur had its problems for would-be breeders. Preparing the pelts was time-consuming and expensive, pushing their cost into the luxury category, while the quality of the furs couldn't compete with mink and chinchilla. During the 1955–56 season, the average pelt sold for a dollar, making nutrias hardly worth trapping, much less feeding and housing. Unlike many other fur-bearing species, nutrias were increasing, not decreasing in the wild, so farming them didn't make much sense. As for pricey mutations, the National Better Business Bureau commented in a cautionary brochure, "It is pertinent at this point to note that a natural brown nutria can be plucked, bleached, and dyed any color of the rainbow for a few dollars."

As long as the hype swelled, nutria breeders could profit by selling nutrias to other entrepreneurs at dream-fueled prices. Some breeding pairs went for as much as twenty-five hundred dollars. It was a grand pyramid scheme, a high-stakes game in which he who took his pelts to market lost.

Twenty years after Ashbrook had sent the nutria pelt to his furrier friend for evaluation, the same questions filled his mailbox. Can you send me information on raising nutria? Where can I get some? How much are pelts worth? Do you have statistics on mutations? Ashbrook dutifully sent a fact-laden and cautionary reply. Finally his measured tone gave way to a burst of frustration. In a pamphlet for the National Better Business Bureau, he wrote:

> The epidemic of booms and busts has been breaking out ever since fur farming began and still persists; silver foxes, muskrats, minks, chinchillas, and nutrias have all experienced a series of these outbreaks. Through the years some legal action has been taken against the unscrupulous promoters of fur animals, and in some cases the fakers have been restrained by court order from staying in business. But by and large they have continued to flourish. Just so long as people are gullible and willing to be victimized, fur farming ranches will continue. It is an appalling fact that in spite of many years of study, research, and promotion of legitimate fur animal production, and utilization of furs, this get-rich-quick skin game can be perpetrated on the American public. Must history always repeat itself?

Eventually the legal action Ashbrook hoped for finally arrived, but the results didn't satisfy him. In 1960 Cabana Nutria ringleaders were indicted for mail fraud. They received less than a slap on the wrist, more like a love pat: suspended sentences and fines of a thousand dollars, less than the price of a single nutria. Cabana Nutria went bankrupt, but a year and a half later the company was back in business—new name, same board of directors.

But by this time the tide was turning. In decades to come a

small but vocal group made inroads into fashion consciousness. Pictures of baby harp seals clubbed to death, blood matting on their white fur, cropped up in magazines. They were soon followed by protesters tossing red paint on the fur coats of matrons attending the opera. Fur never completely disappeared as a status symbol, but it was no longer the ultimate sign of wealth and glamour that it was when Gloria Swanson swathed herself in a chinchilla evening wrap and Marlene Dietrich glowered in a gown trimmed with sable.

As fur declined in popularity, many trappers didn't bother going after nutria anymore. Populations reached unprecedented sizes, and the earlier ambivalence toward the rodents turned to plain dislike. Because nutrias prefer tubers, they rip up the matted roots that support banks and shorelines, promoting erosion and damaging already scarce wetlands. Marsh turns to open water. Riverbanks and beaches slip away. The nutrias also have a taste for young bald cypress, causing swamps of this distinctive tree to vanish. Their increased numbers made the problem only more acute, and the animals, once wooed, were deemed officially unwelcome. In response the Louisiana Department of Wildlife and Fisheries followed in the footsteps of the lamprey eradicators. The state sponsored nutria cookoffs, hoping that a public appetite for nutria chili and nutria sausage would keep the offspring of McIlhenny's imports down to a controllable level. At the request of a local biologist, Chef Paul Prudhomme whipped up a plate of deep-fried nutria to show how tasty it could be. Taking a less lighthearted approach to the problem, the U.S. House of Representatives recently passed legislation that would provide the Blackwater Refuge in Maryland with resources to kill off its nutrias, hoping to erase Dozier and Ashbrook's legacy for good.

Today nutria is still classed as a cheap fur, ranging from $1,995 to $6,000 for a coat, compared with the more upscale chinchilla garments, whose prices run from $30,000 to $100,000. But the modest returns don't stop big dreamers in new areas from pursuing their potential. The lure of easy money in hard times is just too great to resist. While chefs in Louisiana perfect their nutria pâté champagne and officials in Maryland track down the marsh-eating rodents, farmers in Thailand, Slovakia, and other countries waiting for a boom are experimenting with nutria raising, trying to produce larger pelts, angling for durability and shine, hoping for a kind of alchemy that will breed luxury from all their spare dirt.

CHAPTER FIFTEEN

A BORDERLINE CASE

THERE were thousands.

Traveling down the Paraná River of Argentina in a one-masted ship, headed back to meet up with companions on board the *Beagle*, twenty-five-year-old Charles Darwin noted birds with beaks like scissors that skimmed over the water, snapping up fish, and birds with tails like scissors that darted, swallowlike, after insects. On land, jaguar claw marks sank deep into tree trunks. Mosquitoes clustered in dense swarms on any exposed skin. Fireflies blinked in the evenings. In his catalog of everything from evidence of ancient floods to the effect of salt water on beetles, he noted in passing small gray-breasted parrots building nests in tall trees.

It was October, spring in the Southern Hemisphere, and the parakeets would have been preparing for mating season. The birds might have been at work on complex communal nests housing several rooms, each with its own entrance. The male would be collecting thorny twigs for the couple's chamber, snipping them off live bushes, adding an awning, building up the walls. Spotting a particularly fine stick on a neighbor's roof, he might grab it for his own and weave it into the chamber entrance. Inside, the female would have been shredding bits of wood to make a soft lining. If a snake should slither along the branch or a hawk float overhead, alarm calls would rip the air. In the afternoon the birds would gather in a colorful cloud over fields of corn and sunflowers, leaving pock-marked blooms in their wake.

While the young naturalist considered the nest building unusual—all other parrots breed in cavities—and commented on the "great mass of sticks" the birds managed to compile, he didn't dwell on it. Darwin paid more attention to the fact that locals considered the parakeets pests and shot them by the thousands every year.

⚹   ⚹

THERE were a handful.

In the winter of 1970 John Bull, assistant ornithologist for the American Museum of Natural History, had been getting some strange calls. A parakeet was apparently flying freely around New York, defying the snow piled on the ground and the ice slicking the creeks. As reports trickled in, some of sightings as early as 1967 and 1968, he set out to investigate, poking around Valley Stream, Long Island, peering up at the naked tree limbs. And there it was: an unruly ball of sticks among the branches with a perky green bird darting in and out. A second nest bristled nearby.

Bull recognized the stranger as a mid-size South American parakeet, *Myiopsitta monachus*, the only species in its genus. Not as flashy as cockatoos or hyacinth macaws, the birds still make an impression. A green cape starts behind the eyes, covering the back of the head, the wings, and the tail. The tips of the wing feathers are darker, touching blue. The gray feathers spreading over the face and chest give the impression that the bird has dipped its face in shadow and remind people of sobriety and religion. They call it the monk parakeet and the Quaker parakeet.

But surely whoever dreamed up those names was joking. One earful of the parakeets' screams—full-throated, gleeful shrieks— and flashes of Bedlam replace reflections on a monastery.

Bull counted only six birds at the nests, but the wonder was that they were there at all. Rather than lie frozen on the sidewalk, as one might expect, these parakeets repaired their nests, settling

in for the season. Imagining flocks of them winging into corn-fields, he coupled his reports with stern warnings on the potential of the parakeet to prove a monster. Looked at one by one, though, each bird exhibited undeniable attractions. When two fledglings fell from a nest, Bull's wife, Edith, cared for the survivor and named it Charlie.

Over the next few years reports of free-flying monk parakeets piled on Bull's desk. In New York City they fluttered near the cages of captive birds at the Bronx Zoo. Nests reaching the size of picnic tables appeared in broken floodlamps and on utility poles, tucked behind broken window screens. Parakeets turned up near White Plains, at Riker's Island, and in Central Park. Then sightings began to come in from other states. A pair bred in Dallas, Texas. In spring 1971 monk parakeets roosted at a silo north of Lansing, Michigan. In Detroit a stunned bird lover saw two parakeets at his backyard bird feeder in zero degrees. Elsewhere in Michigan they built nests around chimneys to keep warm.

Bird watchers suspected these newcomers were pets, escaped or released. Parrots and parakeets, birds in the order Psittaciformes with their flashy plumage and frequent ability to talk, have been kept as pets for centuries. Natives of the West Indies, Polynesia, and Tahiti brought their pet parrots with them as they traveled from island to island and may have launched parrot populations on islands that didn't have them before. Alexander the Great introduced tame parrots to Europe from the Far East. As they croaked hello in lavish drawing rooms, they became the domestic exotic, a bourgeois vision of foreignness, mocked as pedestrian even in the sixteenth century. To the jaded socialite of Renaissance Europe parrots were no more wild than ostrich feathers or leopard skin print underwear today.

But where had these modern parakeets come from? The answer proved almost too simple. In the United States a window of opportunity opened in the late 1960s and the early 1970s, and the birds flew right through it. In the late 1960s, after scientists had developed antibiotics for psittacosis, a flulike disease that can be passed from birds to humans, imported parrots began pouring into the country. According to records of the U.S. Fish and Wildlife Service, in 1968, 492,280 birds were imported into the United States; 11,745 of those were monk parakeets. In 1971, 27,038 monk parakeets entered the United States, and the total number of exotic birds imported that year reached almost a million. Some of the jump may be attributed to more consistency in filling out newly required forms, but without a doubt the tide of exotic birds flowing into the United States was rising.

By mid-1972, however, the window slammed shut. Fears of Newcastle disease, a contagious avian virus, led to a renewed embargo on imported birds. This was a hindrance, but even so, the desire for the birds continued to grow, and breeders provided what importers couldn't. With legal channels closed, importers tried other ways. In 1973 U.S. Customs Service nabbed a man sneaking across the Mexican border with twenty-four parrots, drugged with tequila and hidden in a box spring in the back of his van.

In the affluent and status-conscious 1980s, parrots and parakeets stayed in the public eye. The birds were gaudy and expensive, appealing to the culture of the times. Magazines pushed tapes that taught parrots to talk, and one psychologist developed a practice specializing in parrots. Pet shops sold to interior decorators searching for just the thing to put a finishing touch on the renovated living room. Between 1982 and 1988 neotropical countries exported 1.8 million parrots; more than 80 percent of them went to the United States.

Then, all of a sudden, pet birds started to appear out-of-doors, on telephone lines, in shrubbery, at backyard bird feeders. No one knew exactly where they came from, but theories abounded. A pet owner tired of all that screeching? A broken crate or sloppy handling of a bird shipment at Kennedy Airport? Importers impatient when they discovered their birds would need to undergo a lengthy quarantine?

A few stories of parrot releases left a paper trail. At the San Diego Zoo officials set a handful of monk parakeets free, imagining they would add color to Balboa Park. Elsewhere in Southern California, owners panicked during a fire and shooed their birds out of their cages in order to save them. In Chicago two double-headed yellow Amazons, brought outside for a block party, escaped when the wind blew over their cage. Hooking up with a flock of blackbirds, the yellow amazons spent weeks shouting Spanish from treetops and telephone lines. *"Hola!"* they screeched to the neighborhood.

But most of the introductions were private affairs. A noisy bird. A cage door ajar. A release.

New York's monk parakeets had flown into a country mulling over the complexities of conservation. In 1973 the Endangered Species Act was fresh legislation, and politicians and importers felt around for the implications. The Wilderness Act of 1964 was less than ten years old; people turned their eyes to millions of acres of roadless areas throughout the United States to determine what was worth saving. Mammals, birds, and even insects and fungi held a new fascination as their fragility was revealed.

More important, parakeets had the misfortune to fly into a world after starlings. They settled into communities much more sensitive to exotic dangers than those of the 1890s, wary of the duplicity of good-looking birds. Results of this deceit showed up everywhere. In 1969, just as the first few monk parakeets struggled through the northeastern weather, the executive director of the Wildlife Society paced on the lawn of the White House, counting bird species, following in the steps of Theodore Roosevelt, who, in 1908, strolled the White House grounds to see how many kinds of birds he could uncover. Among the fifty-six species he tallied, Roosevelt found a golden-crowned kinglet, a saw-whet owl, and more than twenty kinds of warblers. Sixty-one years later the Wildlife Society's man spotted only nine species living near the executive mansion. More than half the birds he counted were starlings. Many of the rest were house sparrows.

As a result, scientists and wildlife managers treated this new wave of settlers cautiously, trying not to do anything rash. What if someone had seen the future of gypsy moths or pigeons and squashed the populations before they got too large? But how could this southern species flourish enough to become a pest in such a northerly habitat? The birds relied on seed at bird feeders to get them through the winter, so maybe they wouldn't venture beyond the suburbs. Experts squinted at the parakeets, attempting to strip away the playful shrieking and the apple green feathers, the quizzical eye and the architectural skill, peering into the boisterous bird soul, asking themselves, "Is this a starling in disguise?"

They decided not to risk it. Representatives of conservation interests in thirteen states met in February 1973 and agreed that the Fish and Wildlife Service should oversee a monk parakeet "retrieval" program. "Retrieval" actually meant "extermination." Monk parakeets' elaborate nests made such easy targets that parakeet hunters could knock them down at will or stake them out and wait for the birds to return home before trapping them with mist nets or offing them with shotguns. By this point monk parakeets had appeared in thirty states, each of which chose to dispatch the invaders in its own way. New York and California, the states with the biggest populations, went after the parakeets with a vengeance and eliminated more than half the birds sighted. On the other hand, Texas and Florida didn't track down any of their parakeets, and Illinois "retrieved" only 10 percent.

These days bird aficionados lacking funds for trips to South America or Australia can see a wide variety of neotropical birds by packing their binoculars and touring North American cities. Canary-winged parakeets nest in exotic palms and flutter through

Miami, stealing mangoes. Brown-throated parakeets send piercing shrieks through the Florida Keys. Yellow-headed parrots pick apart crab apples near New York City and sneak tangerines out of suburban yards in Southern California. Cherry-headed conures clown in San Francisco, swinging upside down from telephone lines and enchanting residents of Telegraph Hill. Many bird watchers by the Bay see the parrots as symbolic of the city itself, where citizens of many countries have chosen to make their home.

That's not all. Orange-chinned parakeets, orange-fronted parakeets, black-hooded parakeets, blossom-headed parakeets, white-fronted Amazons, and orange-winged Amazons all have shown up, some for a few weeks, some for generations. Most can't survive the cold, the lack of traditional food, the alien habitat. But the monk parakeets that eluded the government retrieval program are doing just fine.

⚹ ⚹

THE wrong train deposited me on a weathered station platform whose sagging boards threatened collapse. Stores with broken windows and apartment buildings with no windows at all lined the streets. It's a bitterly cold day, and there does not seem like a less likely place on earth for a tropical bird to roost than the University of Chicago campus when I finally find it. No snow coats the ground, but a fall rain drips and lingers on clothes, against skin. Puddles gather in the corner of cobblestones. Wind ruffles ivy on the walls like fur. No one lingers outside, hustling from dorm to class. It's very far from Argentina.

As I walk out of the visitors' center, repeating new directions to the zoology building under my breath, coffee in two hands, warmth seeping through the paper cup, a high chattering erupts behind me. Dark silhouettes with arrow-sharp wings and long tails dart across the gray sky above the Gothic spires of the campus. A flock of parakeets wheels and heads out toward the water.

In his attic office Professor Stephen Pruett-Jones appears as out of place in this chill medieval haven as do the parakeets. His walls are covered with pictures and stickers of his research specialty, Australian birds with colorful plumage. A border collie–something mix lies flopped under a table. In an adjacent lab *Far Side* cartoons depict the consequences of putting lunch in the same refrigerator as specimens. It's a little sanctuary. Sort of like a nest.

When Pruett-Jones moved to Illinois from California a little more than ten years ago, the flashy monk parakeets were one of the first things he noticed. At first he thought they should be exterminated, but he's grown more ambivalent about these exotics flying around. He still wouldn't mind their being controlled but adds that they don't seem to live anywhere but suburbs or parks or pose a threat to other birds.

"In the vast majority of cases they're totally benign," he comments. "I've never seen a monk parakeet interact with a native species in any way."

With an undergraduate looking for research experience, he began counting them, going out every two weeks, tracking the birds as they sought out dandelions and elm buds in the spring, fruits of ornamental plants in the summer, holly berries in early winter, and birdseed in January and February. As he monitored the Hyde Park population year to year, he could see it was growing rapidly. From April 1992 to March 1993 the number of birds increased by 46 percent, from 64 to 95. By 1997, he estimated, 200 monk parakeets were living in the neighborhood. By now that figure may be up to 250. Nationwide, tallying populations in Florida, California, Texas, . . . numbers are hard to estimate.

"There could be as few as five to six thousand or there could be as many as twenty to twenty-two thousand," Pruett-Jones says.

They have also spread out, reaching into the suburbs of

Bensenville, Berwyn, Calumet Park, and Burnham, somehow persevering through Chicago's legendary cold. The core of birds in Hyde Park is acting as a source population, producing enough parakeets that the young need to abandon the place where they were born and establish new territories. A few have nested in the farther town of Carol Stream, but agricultural fields are still thirty miles away. None has ventured far enough to threaten the rings of grain that radiate from Chicago's center.

While the professor has been tracking the birds for almost a decade, becoming the country's monk parakeet expert, there's much more to learn. Unlike starlings and pigeons, which have been tested and followed, dissected and trained, the monk parakeet is cloaked in secrets. How far do the birds travel to feed? Could they survive without birdseed? What benefits do they gain from living in colonies? Why those gargantuan nests? When there's only one example of an evolutionary trait, tracing its origin can be difficult. There's nothing to compare it with.

Pruett-Jones would like to band the parakeets to track them better, begin to unravel some of the evolutionary questions, and take a stab at predicting whether the populations will explode or settle into a small suburban phenomenon that birdseed can support, but funding is hard to come by. Federal agencies aren't that concerned because the parakeets have not yet caused real harm. So, with a continuing supply of undergraduates looking for interesting fieldwork close to home, he keeps counting and waiting to see what happens.

※   ※

WANDERING through the park on the west side of the university, looking up hopefully at soggy squirrel's nests, I check and recheck the map the professor drew for me of nest sites and good spots for parakeet watching. Cars surge by on a broad avenue. A heavyset woman holds hands with a little girl in a pink jacket and pink hair clips as they wait to cross the busy street dividing one lawn from another. Even in this scrap of green in the city tapestry, it's hard to imagine how a native species could make a living. Roads and sidewalks, office buildings and apartment houses slice up the environment into discrete chunks. Most of the plants are exotic and might not offer much in the way of food to a creature reliant on North American flowers. While I'm hunting for an exotic, how much more startling it would be to see a native species here, one that evolved to make use of plants and habitats nowhere found in this maze of asphalt and landscaping. Then, just past an empty playground, the parakeets have left their mark.

If monk parakeets represent the only parrots to build a nest, they don't take their responsibility lightly. This is no soup bowl of sticks. To call these structures nests stretches the word beyond recognition. One sprawls where a branch forks into three, relaxing as far up one of the prongs as gravity will allow. Another is woven into twigs at the end of a limb, using the living tree as building materials. A third attaches to the tree only at the top and hangs down, swaying in the fall breeze. Messy heaps of twigs, some as big as overstuffed chairs, the nests seem to hold their shape by magic, displaying no structural principle that keeps them from collapsing. Multiple entrances tunnel into the masses, leading to up to twenty separate chambers. In this way they are more like airborne burrows than nests, as if the parakeets were building a new earth, higher than the first, dismissing the pedestrian ground.

This miniworld attracts other animals, just as a house may inadvertently shelter mice and spiders. Ducks lay eggs on top of the structures, as if they were patches of brushy lakeshore. House sparrows nest in the twigs at the edges. In Argentina, when the cavities aren't occupied by parakeets, American kestrels and speckled teals move in. This afternoon, though, the nests appear completely empty and still, like abandoned shacks.

Two days later, still hoping to see parakeets at close range, I head to Hyde Park, on the east side of the university, in the pearl light before sun. Apartment buildings and shops ring the park, and a bus plies a neighborhood street. The few people on the sidewalk seem still wrapped in sleep, brushing aside dream fragments as they move toward their cars. On the neat park lawn, pigeons carpet the grass. Parakeet chatter floats from the trees, but when I look up, starlings fly off, still mimicking their neighbors. Their cries only highlight the stillness of the benches and quiet tennis court. As the professor promised, a huge nest drapes the branches of a tree at the south edge of the park, but in the dim light it looks lifeless.

Then, in a rush, parakeets erupt. Three perch on a sapling by the tennis court, backlit by the sun rising on Lake Michigan. Muttering and squawking, a big parakeet shifts his weight from foot to foot in a crook of a branch above the nest. Every now and then he fluffs up his feathers and runs his beak through them. One flies up carrying a stick twice as long as his body and lands by an entrance. Grasping the stick in his beak, the bird works it into the underside of the nest, twisting and poking until the twig disappears into the mass. Then the parakeet soars off and returns with another prize, continuing renovations. A golden retriever, delirious to be off leash, bounds through the grass, scattering pigeons in a rush. When they settle, two parakeets join them, green among the gray, all pecking at the grass.

For some, these birds are the zany, exotic heart of the neighborhood. Underneath those broad-shouldered suits and midwestern bluster, these Chicagoans are tenderhearted. When the parakeets first started building up a population in the Hyde Park neighborhood in the late seventies and eighties, they were immediately embraced. When an early summer storm uprooted a tree hosting two nests, phone calls asking about the birds' welfare flooded into the Hyde Park Chamber of Commerce. If a nest fell, residents tried to rescue the chicks and find them homes. The former mayor of Chicago Harold Washington treasured the parakeets roosting across the street from his apartment and adopted them as personal mascots. He claimed he could never lose while they nested there. After he died, still in office, newspapers reported that parakeets began building in a tree outside the Harold Washington branch of the DuSable Museum of African-American History.

When the USDA geared up for a second round of parakeet "retrieval" in the late 1980s, Hyde Park residents formed the Harold Washington Memorial Parrot Defense Fund, complete with baseball caps reading "Parrot trooper," and threatened to sue. The birds remained. People still complain when Con Edison, the electric utility, tears down nests built around transformers. The birds seek out the heat-emitting sites, but more than once the stick heaps have burst into flame. As in San Francisco, the birds have been embraced as a symbol of the city's diversity, tolerance, and desirability. People say they are flattered that the birds have chosen their communities, have chosen them.

With the entire Hyde Park population lingering at about 250, there are few enough for them to be pets: pets of blocks, pets of neighborhoods, pets of cities. They rip dandelions from the grass and add to their weighty nests, and I know what I should think. My rational mind urges that they should be exterminated. It's no use waiting until the parakeets start to cause serious damage or until the populations are so big they are impossible to control. But as I watch them swirl the air over the park, I must admit: There is an unmistakable joy in this.

꙳   ꙳

THOUGH parakeets in America's cities may seem bizarre, perhaps the most unlikely of the exotic releases, the years from 1830 to 1970 were the only ones when large numbers of parakeets did not fly freely in the United States. The Carolina parakeet once soared in raucous flocks from Florida to New York, from the southern Great Lakes to Texas, screeching *qui . . . qui . . . qui* and toughing it through some northern winters. The little birds, *Conuropsis carolinensis*, the only ones in their genus, had yellow heads, darkening to orange and rose above the beak, contrasting with green bodies. They nested in cavities of sycamores and cypresses and had a fondness for cockleburs. The surprise of early explorers at finding these birds soaring through the prairies to forage in eleven degrees below zero would be echoed by John Bull seeing the monk parakeets in a New York January centuries later, as would be the tangle of pleasure and annoyance. In 1729 William Byrd complained, "very often, in Autumn, when the Apples begin to ripen, they are visited with numerous flights of paraqueets, that bite all the Fruit to pieces in a moment for the sake of the kernels. . . . They are very Beautiful; but like some other pretty Creatures, are apt to be loud and mischievous."

They even made it to Illinois and flew through the harsh winds that kicked off Lake Michigan. In 1821, when the city in its present form was not even a plan in the mind of an engineer, H. R. Schoolcraft stated: "The paroquet is found as far north as the mouth of the Illinois, and flocks have been occasionally seen as high as Chicago."

But they didn't last long. Soon after it became clear that they skimmed seeds and fruit from fledgling crops, farmers began to kill them. Their bright feathers also attracted hatmakers, market hunters, and sportsmen, who picked them off easily as the birds circled any bird that had been shot. As they grew rarer, naturalists who caught a glimpse shot them to document their find. When it became apparent that the birds were on their way to extinction, Frank Chapman of the American Museum of Natural History traveled to Florida to collect specimens for an exhibit but came back empty-handed. Like the heath hen and the passenger pigeon, by the turn of the century, the Carolina parakeets were as good as gone.

If the biology of the monk parakeet is mysterious, that of the Carolina parakeet is downright cryptic. The last one died in a zoo in September 1914, a bad year for native birds. All that remained to do was to catalog the remains, and only so much can be learned from stuffed skins. Scientists do know the Carolina parakeet was about a foot long, roughly the same size as the monk parakeet, and had a feathered cere (nostril area) and a broad, notched beak, also like the monk parakeet. Moreover, while Pruett-Jones assured me that monk parakeets succeed because they are such generalists, able to live off what they find and not because they are occupying a niche left vacant by the Carolina parakeet, I can't help thinking that the birds shooting like green arrows through the morning to find elm buds in Hyde Park are tracing ancient and ghostly wing patterns in the air.

    *   *

THERE was one.

In the window of the pet store in the Missoula mall, where German shepherd puppies lie on their sides and pant and ferrets are on special, budgies squirm in a glass box. There must be thirty at least—yellow, blue, green, in a sociable muddle. They fan their tails, bang the mirror so it flashes, and gnaw along the edge when it stops swinging. They perch on top of their water dispenser and bounce off and on the square of plastic pipe provided for their amusement. Their ineffectual clipped wings blur, refusing to rise. One jumps vertically, bounding up a foot before plunging back down into the sawdust. Rowdy, like first graders at recess, they lock beaks, seeming to bicker, and chatter away. Kids ignore the sign—THEY WILL BITE!—and shove their fingers through the open top of the compartment toward the budgies, asking to be touched.

In the next glass box over, one monk parakeet paces, picking up chunks of sawdust from the bottom of the box, ripping them to pieces. The tail of this awkward walker swishes back and forth like a sheathed sword. Thin plastic circles, red, yellow, green, and blue, dangle from a climbing structure in the center of the box and colored hard balls lie on the floor. The parakeet ignores them, ignores the sticky hunks of sunflower seeds in the sawdust, fluffs out its gray face feathers, and paces more. Mall shoppers, laden with Gap bags, stop and look. A skateboarder in a cowboy hat pauses by the cage. Of all the parakeets, this one actually seems monklike, standing at the edge of its cube, staring through two panes of glass at the chattering budgies.

This is what people see when they bring a monk parakeet home, all the personality that radiates from an individual. A single bird is so easy to identify with, tell stories about, and name. One cannot imagine killing it, so the owner will open the window and let it go, where it will join others to form a flock, raising wild chicks. And so countrywide, monk parakeets pace the border between individuals and established populations, and no one is sure what the magic number will be and how to calculate it, when they will step across the line that turns pet into pest.

CHAPTER SIXTEEN

A FLY IN EVERY SEED HEAD, A
WEEVIL IN EVERY ROOT

In a small lab in the Bitterroot Valley of western Montana, scientists are fighting a battle with weeds that would give a gardener nightmares. Sulfur cinquefoil, dyer's woad, Dalmatian toadflax, hound's-tongue, beggar's-ticks, leafy spurge, and other invasive introduced plants send down their roots and push through the soil, but spotted knapweed is the worst. More than four million acres in Montana are riddled with the lavender plant that looks like a delicate wildflower to the untrained eye. The trained eye winces. Idaho, Washington, Oregon, North Dakota, South Dakota, and Canada also harbor spotted knapweed, but the problem is most severe in western Montana, the knapweed heartland. Spotted knapweed loves sun-raked, disturbed land, and the Bitterroot offers overgrazed fields, twisting mountain roads, and an increasing patchwork of subdivisions.

The native grasses of these prairies and slopes are bunchgrasses. Rather than grow in the smooth carpets of Kentucky bluegrass or other plants that root themselves with a continuous underground network of rhizomes, they spring up in clumps, like small bushes. Knapweed finds a toehold in the spaces between clumps, sweeping over rangeland once covered by native prairie and turning prime pasture into weed lots. Long taproots let it guzzle water, and its leaves and stem contain a toxin that stunts the growth of nearby plants. Birds and small mammals have a hard time finding cover in knapweed fields. Cows and horses won't eat it. The white-tailed deer and elk that roam wild through this country are reluctant to.

To control invasive weeds, scientists are hearkening back to Riley, Koebele, and their ladybug miracle, putting a modern spin on biological control. Like the cottony cushion scale eating its way through California's orange orchards, knapweed, spurge, and hound's-tongue do so well because they left their natural enemies in the old country when moving on to America. Entomologists aspire to set up a reunion.

In the valley in late May, red-winged blackbirds perch on the barbed-wire fences, then fly off, flashing their epaulets in the sun. A kestrel glides over, red chest, black and white face. A great blue heron flaps toward a rookery on the Bitterroot River, flooding with a rush of spring runoff. Most of the fields are topped with balls of dandelion fluff and carpeted with yellow dandelion sunbursts. Dried knapweed—last year's—clings to the fence along the pasture.

Beyond a sign announcing the "Western Agricultural Research Center, MSU Agricultural Experiment Station," the lab itself is a collection of yellow buildings with brown trim, surrounded by mounds of landscaped flowers. Inside scientist Bill Good and a lab assistant are picking apart knapweed roots, looking for moth larvae. These moths, *Agapeta zoegana*, were brought from Austria and Hungary and released in Montana in 1984 to feed on knapweed. As the lab assistant adjusts the microscope, the radio murmurs country music.

Since knapweed came to the Bitterroot, twelve insect species have been set free in an effort to stop its broad sweep. As a result of biocontrol efforts, Montana now hosts four exotic seed head flies, three seed head weevils, a seed head moth, three root moths, and a root weevil. *Urophora affinis* and *Urophora quadrifasciata*, the most promising of the seed head flies, have flown throughout

the state and are well established. Smaller than a lentil, with tawny striped wings, *Urophora affinis* sets its eggs on the seed head, and the larvae spend the winter in galls grown around them, draining nutrients from the plants and inadvertently providing midwinter snacks for chickadees. Even tinier than *Urophora affinis*, *Urophora quadrifasciata* also lays eggs in the seed head, destroying the floret where the gall forms. The lab raises and ships *Agapeta zoegana*, whose thick white grubs gnaw knapweed roots before turning into bright yellow moths and fluttering away, and *Cyphocleonus achates*, a brown and white speckled weevil, which hollows out taproots when young, stunting the plant and letting in disease, and snacks on tender new leaves as an adult.

In the back of the station is a knapweed field, the center of the lab's distribution operations. Small barriers create a miniature pen for the weevils, which can't fly. Cloaks of netting secure the *Agapeta*, which can. Linda White, a woman with a strawberry blond ponytail, stands next to a box of rocks wearing Tough Guys gloves.

"She's weeding her knapweed patch," Good says.

Though it's not her official title, White is a knapweed farmer. This, in an area where "knapweed" is a worse word even than "Californian," where residents swap herbicide tips over coffee, lowering their voices when they mention some particularly noxious chemical. One day White came home from work to find the census taker. Clipboard in hand, the woman asked White where she worked and what she did. White thought for a moment.

"Well, I transplanted knapweed today," she offered.

The woman just stared.

White raises the knapweed to raise the bugs that she hopes will destroy it. The adults lay eggs during the summer; the larvae spend the winter in the knapweed root. When they emerge in the spring, she sucks the *Agapeta* up with a modified Dustbuster (she collects the weevils by hand), puts them in quart ice-cream containers, and sends them FedEx to county agencies and individuals that want to give the bugs a try. Last year her efforts produced 50,000 *Agapeta* and 12,000 *Cyphocleonus,* which she mailed around the state. Each starter kit contains about 150 *Agapeta* and 50 weevils, enough, she hopes, to launch a war.

꩜   ꩜

WHEN I return a few months later to talk to head scientist Jim Story, the valley has shrugged off spring and embraced summer. The sky holds more than one kind of weather here. Over the Bitterroot Mountains, cut by canyons, dark storm clouds obscure the mountaintops and send an occasional gust to the country below. Up each canyon, the Selway-Bitterroot Wilderness is visible, peaks still snow-covered and forbidding. To the east the Sapphire Mountains appear only as ripples in comparison. In the valley sun presses down on thick air. Sprinklers swoosh over fields, much greener now in July. The river has retreated. The dandelions are gone, but tall purple thistles, relatives of the knapweed, tower over the land surrounding the lab.

In Story's office a large plastic grasshopper peeks down from one of his bookshelves. A paper butterfly rests on the wall. A calendar demands, "Know your category 1 noxious weeds," and offers color pictures of the enemy. Story himself is a gangly man with gray hair and a shirt with big checks. He likes the word "logical" and has spent more than twenty years experimenting with insects that might fancy knapweed.

"If we can tilt the scales a little bit and introduce a natural enemy and put things back in balance, that would seem like a nice logical way to do things," he says.

Story's aim is to get enough species working in tandem to show a significant decrease in knapweed density. He would also like to see the insects in sufficient numbers around the state so that people could collect flies and beetles to bring to their own weed patches. These goals seem achingly modest to those who would like nothing more than to rip the world's last remaining knapweed plant from the earth with their own hands. But Story knows that the plant will never completely be gone and has modified his hopes accordingly. "We're trying to reduce it to a level we can live with," he says.

No one is sure how the first knapweed plant came to Montana. Immigrants from Ukraine may have brought it accidentally in bags of alfalfa seed. Bees that feed on the flowers make notoriously good honey, and rumors speculate that beekeepers moving to the valley spread the plant deliberately to sweeten their arrival in a new place. Regardless, in 1920 the first report of the weed was lodged in Ravalli County, at the center of the Bitterroot. Three years later a few plants cropped up in Bozeman, halfway across the state. By the early 1930s knapweed lined a creek in the Gallatin National Forest, spread through a field at an old fort in Missoula, and grew three feet high along Bitterroot Valley roadsides.

While the initial release of knapweed in Montana is hard to document, the initial release of the insects is easy. Story was there. As a graduate student in the early 1970s he stumbled onto the biocontrol project and knew it was his calling. As he says, "It was a study where I was going to be having to spend a lot of time basically eyeball to eyeball with some insects, which is the kind of stuff I love to do." He introduced *Urophora affinis*, a seed head fly, right here at the experimental station. Soon *Urophora affinis* and *Urophora quadrifasciata*, another seed head fly, had flown throughout the state. The other ten species were released by Story, Good, or both either here at the experimental station or at the Teller Wildlife Refuge down the road.

Dustbusters aside, little has changed in the day-to-day work of biological control since the *Vedalia* cleaned up California: Scientists still travel to foreign countries, hunt for natural enemies of the species they want to get rid of, raise the insects to abundance, set them free, and document the results. But the heirs of Riley, Koebele, and their techniques have also inherited a century's worth of suspicions that introducing exotics can be a very tricky business. "Exotic" and "nonnative" have evolved into cursewords in biological circles, and it takes sweet talking to convince people we might need more of them. While Benjamin Walsh worried he would be laughed at, Story, Good, and White fear unleashing another gypsy moth or Hessian fly. They reassure themselves that in order to get a visa to the United States insects need to pass a series of strict exams. In one test the insects are starved, then offered plant after plant to see if even under extreme conditions they will hold out for knapweed. Any bug that takes the bait is left behind. But it was only recently that scientists began tempting the insects with plants of ecological as well as economical importance. Before this switch the U.S. Department of Agriculture brought over the flower head weevil to devour the European nodding thistle. While it stays away from cash crops like wheat and barley, the weevil has a taste for native thistles, like the Platte thistle, and is severely reducing its numbers. Many don't like thistles, those bristling plants that score the hiker's leg with red lines, so no one is hollering yet to end biocontrol efforts. Still, such tales give Story pause.

"If we ever have a big mistake, it'd shut the whole thing down," he says. But he and the ranchers, farmers, and wildlife managers wrestling with knapweed think it's worth the risk.

Patience. Story, Good, and White have it, and they are hoping others have some too. They have been waiting twenty years for some insect populations to grow large enough to be released, and they are planning to wait twenty more before the knapweed shows signs of thinning out. Ten years ago Story thought he would be able to see knapweed density decreasing at test sites by the year 2000. Now he's readjusting those figures. In the meantime they count larvae in roots. They vacuum moths with Dustbusters. If one insect isn't doing the job, they add another and hope the combination will prove deadly for the plant.

Against leafy spurge, another of Montana's least wanted, bio-

logical control has made significant inroads. The flea beetle *Aph-thona nigriscutis*, a shiny brown insect that looks like a shellacked wood chip, eats spurge leaves and flowers, drops its eggs near spurge roots, and consumes the roots themselves at the larval stage. The beetle eats enough spurge to make a satisfactory difference, but the changes in knapweed aren't as noticeable. Some people's patience is running thin.

The slow pace of biocontrol and the rapid spread of the exotic weeds have caused a rift between environmentalists. In his eloquent book *Grassland*, Richard Manning, a passionate advocate of native grasslands, wants to wrest the remnants of native prairie out of the grip of introduced species, no matter what he has to spray to do it. In an article for *Audubon* magazine, writer Ted Williams observes Hells Canyon blooming with exotics and writes that weeds' "second-best friend is a chemophobic environmentalist." Discussing fungi introduced to combat gypsy moths, biologist Daniel Simberloff writes that we just don't know enough to ensure that biological control won't cause further harm: "Given the poor knowledge of so many insects of non-human-influenced habitats, it is difficult to imagine a protocol that would ensure an adequate scan of regional potential hosts, even aside from the possibility of ecosystem impacts, dispersal out of the region, or evolution." Meanwhile in Missoula, local environmentalists organize protests against spraying for noxious weeds in the Rattlesnake Wilderness Area, citing the herbicide's toxic effects. One woman in the Bitterroot claims to be tracking mutant deer, genetically warped by all the chemicals raining down on the Bitterroot Valley.

Story thinks eventually biocontrol will win out, despite the desires to spray and be done with it. "People are starting to realize that the days of a quick cure are probably gone. That was the nice thing about herbicide. You go out and spray your weed, and the next day the weed's all curved over and dying. It's just not that easy," he says.

In their battle against knapweed, landowners who don't want to spray or introduce more exotic species have few choices. They can bring in goats, which find the weed palatable. They can burn it, douse it with vinegar or hot water, or mow it at strategic times of year. None of these is the cure-all that will put knapweed back in the alfalfa sack. Actually, if any of these methods made a dent, Story, Good, and White would need new jobs.

When one looks at the soft purple glow radiating from the Bitterroot Valley, it's hard to believe that knapweed is disappearing in some places. Yet in its native Eastern Europe the plant is growing more and more scarce. Patches frequented by insect collectors in the past have been swallowed by vineyards and subdivisions. Other collection sites have been turned into parking lots. About the only thing knapweed can't compete with is pavement.

Where the knapweed has disappeared, so have the insects. After one trip Story returned with a handful of *Agapeta*, just twenty-five. Another insect, *Pelochrista medullana*, was approved in 1985. White calls it "the bug of the future," but scientists couldn't find enough of the root moth to ship to the United States. Finally the lab got some eggs six years ago.

Now *Pelochrista* live in a green house in a cage with potted knapweed. Good gave me the tour. Stepping on a stool, he captured a fluttering moth in a petri dish and pointed out the swelling abdomen that indicated it was female. They are trying to get a greenhouse colony started, but it's slow going. The insects hatch at the wrong time of year because of the greenhouse conditions, and the scientists may see a yield of only three adults from fifty larvae. But these few pots of knapweed may be more food than *Pelochrista* can find in Europe. This cageful may be the biggest population of the moth in the world.

Just as abundance makes knapweed a nuisance, scarcity increases its charm. In Candida Lycett-Green's book *England: Travels through an Unwrecked Landscape*, she describes glorious banks in Gloucestershire that "brim with meadow crane's bill, hogweed, yarrow, lady's bedstraw, scabious, knapweed, and sudden patches of rosebay willowherb." Admittedly there are many kinds of knapweed, and I don't know which the author is describing. But it's hard to imagine Bitterroot Valley residents lured to vacation in England by this description.

⅓   ⅓

A TRIP through Montana offers a glimpse of a state locked in a battle with exotic plants. Looking at the grasslands rippling down from mountain peaks and rivers slicing through rock and losing themselves to meadows, landscapes that would pass for wilderness anywhere else, one might ask why. The answer would be "Because there's so much to lose."

An hour north of Corvallis, along the banks of the Clark Fork River as it cuts through Missoula, the knapweed is blooming. It has claimed this north side of the river, creating a hedge only interrupted by rocks and cottonwoods. The weed lines pathways like this popular riverfront trail, tracing the course of people, cattle, and cars. During the spring the white dry husks emerged from under the snow and lay crouched, bleached, and frail as old mouse bones. Then the new green shoots pushed up through the ground and only recently burst into lilac. Honeybees land and fumble for nectar. Pale green stalks and leaves with gray-brown seed heads look fuzzy, feel prickly. Knapweed is related to bachelor's buttons, though it has the scraggly, tough attitude of an uncultured plant. Here, in the early summer, some of the swollen heads are tipped with purple. Others, open further, offer a brush dipped in lavender. It's a composite flower, like a daisy, made of tiny flowers clustered in the center with white tongues and outside petals that start as a spike and burst into five parts, a small, fragile claw. The base is filled with seeds. The small leaves spring from a stem tough enough to cause a bystander to break a sweat trying to uproot one or merely to pluck a blossom. Despite it all, the air smells sweet.

Across the river the bank displays a different character. In this spot the Clark Fork is paralleled by a slower side stream, home to muskrats and claims of sightings of hundred-pound beavers. In between the mowed lawns and landscaped mounds of the *Missoulian* newspaper offices and the rest rooms adjoining a football field, a patch of native prairie has been restored. Clarkia grows here. So do shooting star, bitterroot, and bluebunch wheatgrass. This is a mixed polyculture rather than a monoculture along the river, a variety of plants, the most noticeable of which is the wheatgrass, shooting skyward in tufts. Here bees are also busy, and grasshoppers bounce through the stalks, but it's only a small patch. One bison would make short work of it.

Even farther north, in Glacier National Park, knapweed drops seeds along the roadsides. At the bases of jagged, sky-scratching peaks and in the glacier-scooped valleys, hotels and campgrounds establish a human stronghold in the land of grizzlies, wolves, and mountain goats. Each rest area has a parking lot, and in the spots of grass near these parking lots, knapweed waves a few purple pom-poms of victory. Most of Glacier's backcountry is still dominated by the vivid red, blue, and yellow of paintbrushes, lupines, and glacial lilies, but a few bursts of lavender have made it beyond the pavement edge to remote streams and lakeshores.

The seeds hitch rides on tourists' cars and hide in shipments from contractors. In an effort to retain the traditional ecology of the park, purification has become an obsession. Officials inspect all gravel and dirt brought into the park and contractors must wash their vehicles before entering. No hay is allowed in the backcountry, and hay that crossed park boundaries must be certified as "weed seed free." Despite these efforts, some weeds have taken root. While the parks have a mandate to keep exotic species at a minimum, Glacier uses introduced insects for both spurge and knapweed. With more than a million visitors to the park every year, each with the potential of carrying seeds on wheels or boots, the choice becomes not whether exotics but which.

On the other side of the Rocky Mountain Front, where the mountains collapse into the plains, lies a rare water-rich area called the Pine Butte Swamp. The preserve, managed by the Nature Conservancy, contains the largest fen in the Rocky Mountains. In spring, when hundred-mile-an-hour winds race through the grasses, grizzly bears stumble out of mountain hibernation down to the plains, lured by the lush stream-side habitat. Today the winds are only blustering, and the grizzlies are out of sight, but a white pelican flaps over the fens, sharp-edged against the dark water. About thirty miles away three flat-topped buttes heave off the level ground, but flatness wins out in the far distance, as blue sky bleaches out to white.

Here too knapweed and leafy spurge poke through the soil. Managers, worried that the more than seven hundred species of plants at Pine Butte will turn to a knapweed monoculture, spray with Tordon and 2, 4-D. They walk the preserve, scanning the ground for new starts. The Nature Conservancy, like the National Park Service, is hesitant to introduce more exotics, but the ranchers on all sides use the insects, so infestation just seems like a matter of time. Pine Butte is the first Nature Conservancy property to permit biocontrol. Somewhere, down where the shadows of clouds slip over the prairie, the flea beetle gnaws at leafy spurge.

Dave Carr, Pine Butte's manager, surveys the scene and explains the contradiction. "There's a risk if I do it, but if I don't do it, we're going to lose the whole shebang," he says.

Farther east, rancher Robert E. Lee speaks of grass with a religious fervor. The Lees raise cattle, wheat, and barley on a 880-acre ranch near Judith Gap, in the plains of central Montana. Grass, how it fares through the winter, how early it shoots up in spring, how much meat each variety adds to his cows: All these are Lee's passion. Noticing a patch of grasses by the road, he can't resist crouching down to examine it, running his fingers over the stalks, pulling up a shoot, and chewing on it.

His biggest fear is exotic weeds. The Lees spend a lot of money on weed control, and Kathy, Robert's wife, spends a lot of time walking the roads along their property, checking for knap-weed shoots. A few leafy spurge plants have taken root, and he's introduced the flea beetle to go after them. He shivers when he thinks of the Bitterroot Valley and all its knapweed. While it's hard to picture a rancher intimidated by a little purple flower, the thought of those lavender plants pushing up through his grass preys on Lee's mind. "That just scares me, I can't tell you how bad that scares me," he says.

Like Jim Story, Lee isn't opposed to exotics per se, just the ones he thinks are damaging. His cows are exotics after all. Each year he plants a pasture of crested wheatgrass that pokes above the snow before the native grasses, providing his cattle with early spring forage. Despite the benefits for ranchers, studies have sug-gested that crested wheatgrass might cause more soil erosion and provide fewer soil nutrients than native grasses.

Back in Missoula at an old fort along the Bitterroot River, in a field abandoned for years and grown into a knapweed thicket, Cub Scouts and Girl Scouts kneel in a cleared patch of earth, uni-forms splotched with dirt, sweat leaving muddy tracks on their faces, quietly bickering over the trowels. Part of a native prairie restoration project, they scrape pockets of soil, shake tiny starts of bunchgrasses and wildflowers from planters to loosen the roots, and press them into the ground, evenly spaced. Heaps of three-inch-high knapweed plants, pulled out by hand, line the edges of the plots, waiting to be tossed into big black garbage bags and carted away. Unlike many of their ancestors or Sheldon Jackson, who saw a waste of plains crying to be converted to productivity, they know the prairie plant names—june grass, bluebunch wheatgrass, hairy golden aster, prairie coneflower—see their value, and are willing to spend this broiling afternoon trying to bring them back. It's risky. Knapweed, hound's-tongue, and Canada thistle creep along the riverbank nearby, waiting for the chance to retake the field. The deer will nibble at the new shoots that very night. It might not work, and if it does, it will be slowly.

But, as Story says, "There's no perfect answer, but we're trying to do the most logical thing we can."

CHAPTER SEVENTEEN

NARROW MISS

As gypsy moths leave another tree naked in spring, star-lings boot cavity nesters out of every desirable hole, and sea lam-preys eviscerate a final Great Lakes trout, it's easy to overlook the fact that things could be worse. It's easy to forget to be thankful for all those exotic species that died en route, couldn't adapt, failed to breed, or, for whatever reason, just never quite made it.

The kangaroo, for instance.

In 1892 Robert C. Auld, a cattle breeder originally from Scot-land but relocated to the Midwest, noticed the buffalo were grow-ing scarce. With the herds gone, vast reaches of prairie stood empty; acres of grassland—wasted. What the West needed, he suggested, was a new large ruminant to take its place. A beast to graze, to convert all that grass to cash, to offer amusement in those long prairie days. The kangaroo would be perfect.

Auld considered all the angles. From shapely ear to hefty hind leg, the kangaroo exuded potential. Hides went for five to seven dollars a dozen. The fur wrapped the shoulders of the stylish. The tails made tasty soup. Sportsmen could save the cost of the fare to Australia and cull trophies from their own backyards. But his vision didn't stop with the creature as an expensive curiosity. "There may come a time when it may become more profitable to raise kangaroo than even cattle on the 'arid' ranches," he wrote.

His research revealed that kangaroos had been released in 1838 in Aberdeenshire, Scotland (the introduction didn't take for they all were the same sex). A second population, also imported to Scotland and placed near railroad tracks, used to delight those on the train who passed by them.

Australian connections assured Auld that the operation would be easy and cheap. They suggested the great gray kangaroo and the swamp wallaby for maximum profit and noted the kangaroo could eat whatever cows disdained and should have no trouble with the North American climate. Both males and

females were reputedly docile and easy to handle. The only expense would be fences. High ones, of course.

＊ ＊

THAT was a narrow miss. But it's easy to picture what might have happened. The story is sadly too familiar. It could have been something like this:

The first kangaroos were welcomed, greeted at railroad stations by banner newspaper headlines and nervous professors reading long speeches. As they hopped tentatively from their crates, the crowd sent up a cheer. The fashion that year demanded mothers carry their babies in beige dresses with crocheted front pouches. Towns renamed themselves: Kanga City, Joey, and Rooville. Fortune hunters sank life savings into buying one or two breeding pairs.

After three months, though, the blush was off the rose. One swift kick or high leap dispatched the ranchers' fences, and the kangaroos roamed the plains in feral bands, harassing the antelopes. They proved more fruitful and hungry than Auld imagined, turning lush fields into stubble. Elk that shared the same habitat began to look a little thin. Dresses with pouches flooded thrift stores. Rooville voted to go back to being Copper Town. All the leggy young men hired on as kangaroo ropers drifted off in search of other work. Meanwhile the kangaroos ate and bred and hopped and proved what Australian sheep farmers always knew: They're pests.

One day, when kangaroos had become common as grasshoppers, decades after people forgot they were nonnative, news arrived that shocked the region. A reclusive botanist had sent pages of research notes written out in longhand to a premier scientific journal, observations based on years of wandering meadows where the kangaroos grazed. Seeds of the rare coulee-dwelling trembling wallflower proved unable to germinate in the compact soil resulting from the bounding herds. The few that did struggle through the earth were quickly chewed up as the exotics found the young shoots a delightful snack. The species perched on the edge of extinction.

Then the real controversy began. Kangaroo enthusiasts who'd grown up loving the gentle hop, hop, hop of the animals as they nibbled tender broccoli buds from the garden at dusk sent haiku and Petrarchan sonnets to their governors. Father and son duos offered gruff testimony about the thrill of hearing the boomer's warning thumps at their approach and the joy of celebrating a successful hunt with savory kangaroo medallions. The bartender at Hopper's and the owner of the microbrewery Big Tail Pale Ale overcame their differences and cowrote an impassioned letter to the editor about the shortage of good mascots these days. Meanwhile the antikangaroo factions declared them noxious weeds, refused to refer to them by any other term than "the invaders," and demanded every last one be shot (and drawn and quartered for good measure).

State and local legislatures earmarked hundreds of thousands of dollars for kangaroo control measures. A brave zoologist ventured into the field with ten thousand packets of birth control pills and a prepared explanation, never to be heard from again. When government scientists brought the dingo over as a biological control agent, voles, marmots, ferrets, snowshoe hares, and miniature dachshunds vanished from Sioux City to Spokane. Still, the kangaroo numbers continued to grow. In the swell of a population explosion, with the range grasses depleted, they stumbled into towns only to find kangaroo-proof AstroTurf lawns, installed by federal mandate. There they collapsed on the sidewalk, where the postmaster tripped on the destitute creatures. It would have been ugly, no doubt.

*   *

BUT despite the solid reasoning of Auld's plan, no popular upsurge demanded wallabies for Wyoming. Maybe funds were short. Maybe will was lacking. Maybe it just wasn't the right time. Eventually Auld turned away from ranching and moved to New York to become a newspaper journalist. In later life he studied up on the Scottish poet Robert Burns, wrote a book called *The*

Robert Burns We Love, and served as founder and general secretary of the Robert Burns Memorial Association. The kangaroo scheme fell by the wayside.

So even though the country may seem overrun, invaders pawing at the gates and taking root under them, rest assured that it could have been worse. Envision the kangaroo herds bounding over the horizon, headed for downtown. Then send up a small prayer of gratitude for whatever quirk of fate kept the prairies kangaroo-free. At least for the time being.

AFTERWORD

AT the Smithsonian Museum of Natural History, exotic species are on display. On the sidewalk, in front of the museum, facing the Mall where the Capitol building presides at one end and the Lincoln Memorial at the other, pigeons pursue one another, males inhaling to fill their feathered chests, exhaling a series of coos. The females nod and bob, choosing from potato chips and bread crusts left by students when the field trip was over. Starlings are strewn in front of the Washington Monument, sleek feathers and speckles bold against the freshly cut grass. In the butterfly garden, a bank of flowers on one side of the museum planted to tempt native butterflies with their nectar, honeybees steal pollen. They strip it out of the flowers' centers with their jaws and comb it off their bodies, packing it into pollen baskets.

Inside, upstairs in an insect exhibit, a whole hive raises young, cures honey, and builds up the waxy comb, all behind glass, exposed to visitors' eyes. The hive is connected to the outside world by a Plexiglas tube that penetrates the museum wall. Bees fly into the tube, probably weighted with pollen from the butterfly garden, and waggle-dance on the surface of the comb, stepping over one another, shaking.

Downstairs, enclosed in a glass diorama, a life-size mannequin of John Smith, wearing a metal helmet, receives corn from life-size mannequins of Powhatans. A interpretive sign explains how hungry the colonists were when they first arrived. Here, in the nation's capital, the two histories breathe side by side.

It is tempting, as one looks at this scene, to wish it all away. To conjure, instead of the familiar countryside this book opened with, the landscape that met the first European explorers as they eased their ships up to the beach. A forest awake, stirring with the noise of thousands of passenger pigeons roosting in beech trees, rustling the branches, cooing to one another. Lake trout filling the Great Lakes and westslope cutthroat healthy on the west slope of the Rockies. The Olympic Peninsula pushing up towering spruce trees with timber wolves skirting their bases. Salmon, surging upstream, all slick scales and instinct, back after back, leaving hardly any room for the water.

Compared with this vision, these tales of exotic species are steeped in sadness. While they appear tales of addition, subtraction is the underlying theme. The accounting is unforgiving. Instead of a world with passenger pigeons and rock doves, we have only rock doves. Instead of a world with heath hens and ring-necked pheasants, we have only ring-necked pheasants. Instead of a world with Carolina parakeets and monk parakeets, we have only monk parakeets. In this light, the desire rises to see the country before the first European bootprint in the mud, showing off the creatures it molded, the children of its brain and habitats. It would be frightening; it would be awe-inspiring; it would be a place we've never been.

Or ever will be. It's not chance that honeybees, pheasants, English sparrows, and brown trout do so well. They know how to live with us, biologically understand agriculture, angling, and, in some cases, cityscapes. We create more habitat for them all the time.

But that's not to say we should give up, either. Government regulations—the ones that govern the dumping of ballast water and the importation of the millions of plants and animals and insects that pour into the country every year—need to be strengthened. Biologist Daniel Simberloff suggests replacing the current system of blacklisting imported species—in which a species has to prove it's harmful before it's banned—with a "whitelist," in which species have to prove they're safe before they can cross into the country. Even with the current system, blacklists need to be updated more quickly to save the poor botanists the heart-stopping anxiety of flipping through a nursery catalog and seeing the exotic plant that is currently encroaching into their favorite wetland, grassland, or hillside attractively photographed and for sale.

Once a species makes it into the country, quick action can stop it. This relies on a willingness of state and federal agencies to recognize potential problems and heave money and resources in their direction before they balloon out of control. Harder still, it depends on the ability of gardeners, hikers, and land managers to recognize what's native and what's not, what has been around and what's new, to cultivate a sense of biological history.

Restoration is a growing field, and scientists and managers constantly learn more about this delicate art. While many of the projects center on plants (though the highly publicized reintroduction of wolves may be counted as restoration of a sort), these reclaimed communities provide valuable habitat for the species from bees to grizzly bears that live there. A National Wildlife Federation program trains people to become habitat stewards, planting flower gardens, schoolyards, and even plots of land near businesses with native plants, offering patches of habitat interwoven with the city matrix. Larger-scale efforts are also under way, often fueled by volunteers who devote long hours to pacing an island or scouring a canyon they've adopted, looking for and ripping out any hound's-tongue, tamarisk, or yellow star thistle that they see. Reclaiming the Garden, it appears, involves a lot of weeding.

But with all these fixes, the hardest problem lingers: our restless nature. What makes the disasters springing from these exotic

species introductions so perplexing is that at times they spiral out from our best impulses: an urge to feed the hungry, a care for the Olympic Mountains and their rain-sodden valleys, a plan to help farmers grow healthy orange orchards, the need to develop a cure for polio or malaria, a love of Shakespeare. Many of the motivations spring from traits that make us proud to be human, even though we look back and find the actions inconceivable.

And on it goes. New frontiers open all the time. One of the most alluring is the miniature universe of DNA, where even now tinkerers are hard at work, trying to improve what they find. Only recently, altering the genetic structure of corn to produce a strain that could shrug off insect attacks, researchers came up with a type they thought would work. It also, as scientists at Cornell discovered, killed the caterpillars of monarch butterflies that fed on milkweed leaves sprinkled with pollen from the improved corn.

So we should do what we can, take actions that make the most sense to us given our present understanding, proceed with caution, work to expand our peripheral vision so it takes in more species and unglimpsed possibilities, reach to see beyond the effects we hope to achieve. We should also rest assured that in the half-light at the end of the working day, no matter how wide we open our eyes and how finely we tune our fortune-telling instruments, no matter how many times we recheck the calculations and stretch to account for the earth complete and entire, the natural world will continue to rattle, buck, elude, and astonish us, serving up results far beyond the imagination.

ENDNOTES

INTRODUCTION

3   The statistic that 99 percent of the biomass of the San Francisco Bay is exotic comes from Cohen and Carlton (1998), p. 556.

3   The statistic that exotics may have contributed to the decline of 49 percent of threatened and endangered species comes from Wilcove et al. (1998), p. 607.

4   The statistic that more than forty-five hundred exotic species live in the United States comes from Onstad and McManus (1996), p. 430.

CHAPTER ONE

12   "We had no other domestic . . ." Lescarbot (1968), p. 226.

12   "The savages had no knowledge . . ." Ibid., p. 227.

14n "One relic of the good . . ." "Life in Brittany" (1880), p. 131.

14   "induced me to expose . . ." Champlain (1907), p. 17.

15   "Take peions and stop hem . . ." *The Forme of cury,* quoted in *Oxford English Dictionary,* vol. 7 (1961), p. 8450.

15   "The Spirit cam doun . . ." John Wyclif, *Selected Works,* I, quoted in *Oxford English Dictionary,* vol. 3 (1961), p. 621.

16   "She is coming, my dove . . ." Tennyson (1899), p. 252.

18   "Into the breach went . . ." J. L. Carney, quoted in Levi (1974), p. 10.

20   "not infrequently lingers too long . . ." Townsend (1915), p. 314.

20   The statistic about the increase in feral pigeon numbers in El Paso between 1972 and 1992 comes from Johnson and Janiga (1995), p. 232.

22   "Among all the most useful . . ." Champlain (1907), p. 17.

CHAPTER TWO

26   "what a paradise this . . ." Smith (1907), p. 201.
26   "a Country which nothing but ignoarance . . ." Kingsbury (1933), p. 545.
26   "These waters wash from the rocks . . ." Smith (1907), p. 82.
27   "the purling Springs and wanton . . ." Gent (1963), p. 27.
28   "these few private lines . . ." Kingsbury (1933), p. 534.
29   The statistics on the number of colonists sent over by the Virginia Company and the number that survived are from ibid., pp. 537, 551.
30   "We have by this Shipp . . ." Ibid., p. 532.
30   The statistics on the number of colonists sent on the *Bona Nova, Hopewell,* and *Discovery* are from ibid., p. 639.
30   "open . . . hives and give Bees . . ." Gervase Markham, excerpted in Rasmussen (1960), p. 9.
31   "if men would endeavor . . ." "A New Description of Virginia," in Force (1963), p. 15.
31   "Nature is so great . . ." Pliny (1991), p. 149.
31   "I'll tell of a tiny . . ." Virgil (1957), p. 77.
31   "[Bees] will have the power . . ." Montaigne (1927), p. 469.
33   "Now if the facts are so . . ." Arthur Dobbs, quoted in Proctor, Yeo, and Lack (1996), p. 16.
34   "the white man's fly." Jefferson (1975), p. 111.
36   The statistic that four-fifths of commercial crops in the United States are pollinated by European honeybees comes from Buchmann (1996), p. 194.

CHAPTER THREE

42   "large armies of foreign mercenaries . . ." Jefferson (1984), p. 21.
42   "[T]hey should have kept . . ." Martin (1962), p. 121.
43   "I would beg leave to propose . . ." *Pennsylvania Gazette,* quoted in "Newsclips of Sorcerers, Pests, and Downright Ugly Matters in 1787" (1987), p. 38.
43   "manifest tokens of the displeasure . . ." *The Prophet Nathan, or, Plain Friend* (1788), p. 13.
43   "They have fallen almost . . ." Ibid., p. 10.
44   "Had a company of soldiers . . ." Asa Fitch, quoted in Wagner (1883), p. 31.
44   The suggestion that the Hessian fly be called wheat destroyer is from Wagner (1883), p. 28.
45   "Three Hessian flies only were seen . . ." Conway (1862), p. 148.
45   "Patriotic motives are the worst . . ." Hagen (1883), p. 45.
45   "I consider, therefore, the Hessian fly . . ." Ibid., p. 49.

CHAPTER FOUR

49   The statistics on the changes in the population of the Hawaiian Islands between the time of Captain Cook's arrival and Dr. Judd's come from Morgan (1948), p. 114.

49   "O the mosquitoes! . . ." Judd (1960), p. 66.

49   The statistics on the number of whalers docking at Lahaina in 1826 come from Morgan (1948), p. 78.

53   "A kind greeting, a shower bath . . ." Lyman (1925), p. 120.

53   "The bad and the good . . ." "Social Life in the Tropics" (1868), p. 567.

56   "When I first arrived in Kona . . ." R. C. L. Perkins, quoted in Munro (1960), p. 69.

57   "It is more reasonable . . ." H. W. Henshaw, quoted in Warner (1968), p. 102.

61   "the thing with feathers . . ." Dickinson (1960), p. 116.

CHAPTER FIVE

64   "I felt as if approaching . . ." Thomas Moore, quoted in Dow (1921), p. 131.

64   "Niagara was at once stamped . . ." Charles Dickens, quoted in Dow (1921), p. 230.

65   "Niagara Falls . . . is tremendously high . . ." Sebastian Vauban, quoted in Jackson (1997), pp. 26–27.

65   "These seas affording . . ." W. H. Merritt, quoted in St. Catherine's Historical Museum, p. 6.

66   "The artificial wedding of the Great Lakes . . ." W. H. Merritt, quoted in Jackson (1997), p. 42.

70   The statistics on the trout catch in Lake Michigan in 1945 compared to 1949 come from Munro (1950), p. 138.

70   The statistics on the change in lake trout population in Lake Huron over the course of twelve years come from East (1949), p. 424.

71   The statistics on the number of sea lampreys currently in the Great Lakes come from Mahan and Mahan (1998), p. 76.

73   "Against immortal foes they stood . . ." Wilson (1825), p. 2.

CHAPTER SIX

80   "The labor of a few old men . . ." Andrews (1871), p. 181.

82   "so much prospective importance . . ." Ibid., p. 181.

85   "What a destruction of leaves . . ." Trouvelot (1867), p. 85.

85   "In their season I used . . ." Forbush and Fernald (1896), p. 8.

86   "They were all over the inside . . ." Ibid., p. 8.
86   "The caterpillars would get into . . ." Ibid., p. 9.
86   "I spent much time . . ." Ibid., pp. 9–10.
86   "the clipping of scissors" Ibid., p. 18.
86   "a breeze" Ibid., p. 20.
86   "the pattering of very fine rain drops" Ibid., p. 16.
87   "In one of these drawings . . ." "Astronomical Plates" (1872), p. 4.
88   The observation that the gypsy moth was originally misclassified in the same genus as the silkworm comes from Hubbell (1996), p. 140.

CHAPTER SEVEN

91   "To the day we celebrate" Scrapbook 108, p. 87.
92   "On one occasion . . ." Owen Denny, quoted in Shaw (1908), p. 12.
96   "Ten years from now . . ." "Paper Shells" (1882), p. 1.
97   "They fly swiftly, run fast . . ." Averill (1895), p. 56.
97   "It is not too much . . ." Jennie Griffith, quoted in McGuire (1899), p. 148.
98   "This is the sport of civilization . . ." Sears (1897), p. 983.
99   "There are a kind of fowles . . ." Morton (1883), pp. 193–94.
99   "So numerous were they . . ." Elisha Lewis, quoted in Gross (1928), pp. 521–522.
100 "[O]ur local sportsmen . . ." Greene (1893), p. 385.
100 "Men, still young today . . ." Shaw (1908), p. 17.

CHAPTER EIGHT

106 "We should hail the day . . ." German Society of Fish Breeders (1880), p. 911.
108 "No care or labor . . ." Stone (1876), p. 390.
108 "a strange, new business . . ." Mather (1873a), p. 10.
108 "As the Hudson River . . ." Ibid.
110 "The American department was a complete . . ." Haack (1882), p. 57.
111 The statistic on the numbers of brown trout eggs in each shipment comes from Mather (1886a), pp. 143–44.
112 "This fish seems given . . ." Mather (1886b), p. 133.

CHAPTER NINE

118 "Methought he ought . . ." Asa Fitch, quoted in Barnes (1988), p. 67. Like Walsh, Fitch noticed that natural predators kept species in check in their native lands and mulled over the idea of biological control.

119 "The simplicity and comparative cheapness . . ." Benjamin Walsh, quoted in Doutt (1973), p. 30.

120 The statistic that almost six hundred species and varieties of trees were imported into California between 1810 and 1942 comes from Stoll (1995), p. 216.

123 "He was the best-loved . . ." Smith (1910), p. 476.

124 "I have not yet met anybody . . ." Letter from C. V. Riley to Albert Koebele, October 3, 1885, Koebele Collection, Box No. 1., California Academy of Sciences Archives.

127 "Your bugs are all right." Koebele (1890), p. 28.

127 "the dry bodies of the *Icerya* . . ." Coquillett (1889), p. 74.

130 "the Koebele method" Howard (1930), p. 502.

130 "[T]o Koebele alone is due . . ." Scrapbook, Koebele Collection, Box No. 3, California Academy of Sciences Archives, p. 23.

132 The statistic that Frank Meyer introduced twenty-five hundred new plant species to the United States comes from Manning (1995), p. 172.

132 "I feel ashamed to eat . . ." Howard (1925), p. 560.

CHAPTER TEN

135 The information that Schieffelin released eighty starlings on March 6, 1890, comes from Chapman (1906), p. 164.

136 Details of the social situation in 1890 come from the *New York Times* for 1890. Information about the weather and the resulting sledding through the streets comes from the issues for March 6 and 7, 1890.

137 "the introduction and acclimatization . . ." American Acclimatization Society (1871), p. 3.

138 "We hear the note . . ." Bryant (1910), p. 373.

138 The information about the release of starlings in New Jersey in 1844 and Oregon in 1889 comes from Long (1981), p. 360.

139 The information that one thousand house sparrows were introduced into Philadelphia by city officials is from ibid., p. 375.

139 "Only seven out of the 24 . . ." John Burroughs, quoted in Barrus (1925), p. 216.

139 "He said he would not ransom . . ." Shakespeare (1974), p. 853.

142 "a starling which do whistle . . ." Samuel Pepys, *Diary I,* quoted in Simpson and Weiner (1989), vol. 16, p. 536.

142 Details about the spread of the starling through New York City come from Chapman (1906), p. 164.

142 The statistic about the current number of starlings in the United States comes from Page (1990), p. 76.

142 "From the bird-lover's point of view . . ." Chapman (1906), pp. 164–65.

143 "The Poet may sing . . ." Mather (1881), p. 46.

143   Information on T. S. Palmer's paper about the dangers of introducing exotics comes from Allen (1900), pp. 396–97.

145   Details of Norman Weitzel's study of the starlings in his cottonwoods comes from Weitzel (1987), pp. 515–17.

CHAPTER ELEVEN

149   "In 1881, when I first . . ." Woolfe (1891), p. 7.

149   "The orphan children . . ." W. E. Roscue to W. J. Harris, October 7, 1890. National Archives.

153   "preaching the glacial gospel . . ." Muir (1993), p. 24.

155   "[I]t has been seriously represented . . ." "Reindeer for Alaska" (1892), p. 4.

155   "those vast, dreary, desolate . . ." Jackson (1891), p. 4.

155   "In our management of these . . ." Jackson (1893), p. 34.

156   "the poor Eskimeaux" These references appear frequently in letters to Jackson in the National Archives.

157   "they have no knowledge . . ." Jackson (1893), p. 9.

159   "It seemed as if . . ." W. T. Lopp, quoted in Jackson (1894), p. 119.

159   "It occurred to me . . ." Miner Bruce, quoted in Jackson (1894), p. 64.

162   "Good pasturage for reindeer" Ibid., between p. 16 and p. 17.

162   "I can not believe Doctor Jackson . . ." Churchill (1906), p. 116.

162   "I have been informed . . ." Ibid., pp. 116–17.

162   Statistics from the 1932 reindeer roundup on the Seward Peninsula come from U.S. Department of the Interior (1933), p. 13.

163   "Well, there is no profit . . ." "Statement of L. J. Palmer, after being duly sworn, taken at Fairbanks, AK in the presence of C. R. Trowbridge and H. M. Gillman Jr. Field Representatives on the 24th day of May, 1932." Smithsonian Archives.

163   The statistic that reindeer populations reached 650,000 in the 1930s comes from Leopold and Darling (1953), p. 69.

164   The statistic that in 1950 the population crashed, leaving only twenty-five thousand reindeer in Alaska comes from ibid., p. 71.

164   "By 1938, it included . . ." Scheffer (1951), p. 356.

165   Information about current herding of twenty-five thousand reindeer as part of a $1.6 million per year business comes from Mitchell (1996).

CHAPTER TWELVE

170   Specifics about the social and economic situation at the time were taken from issues of the *Seattle Times* covering late 1924 and early 1925.

171 "He wonders too . . ." Webster (1921), p. 50.

176 "He looked white and huge . . ." Wister (1893), p. 41.

176 "I like to think of him . . ." Singer (1925), p. 313.

176 "as an animal to whom . . ." William T. Hornaday, quoted in Webster (1920), p. 146.

178 The statistic that in the 1970s the goat population was increasing at 20 percent a year comes from Olympic National Park (1987), p. 20.

178 The estimate that there were twelve hundred goats in Olympic National Park in the mid-1980s comes from ibid., p. 1.

178 "Management of populations of exotic . . ." National Park Service (1994), p. 6.

178 "provide permanent protection . . ." Ibid., p. 7.

180 The statistic that the park service removed six hundred goats in the late 1970s and 1980s comes from ibid., p. 2.

180 The information that some of the captured females were lactating nannies comes from Wagenvoord (1995), p. 35.

180 The information that the goat population dropped to 250 in the mid-1990s comes from ibid., p. 33.

CHAPTER THIRTEEN

184 "Live Longer and Better," "Empire of Sunshine," and other information about the way Florida publicized itself comes from travel advertisements in the *New York Times* during the late 1930s.

186 "These lands are now . . ." Hallock (1876), p. 246.

187 "South-Sea atmosphere in Florida" Ad for Fort Myers, *New York Times*, December 11, 1938, section 11, p. 9.

187 "See the most amazing spectacle . . ." Ad for McKee Jungle Gardens, from ibid., p. 8.

189 "Tarzan couldn't tell it . . ." Putnam (1941), p. 41.

191 The statistic about the number of rhesus monkeys imported into the United States in 1938 comes from Sanders (1940), p. 284.

194 "untouched and untamed" "Lost River Voyage" (1999).

CHAPTER FOURTEEN

198 "Her deep lustrous eyes . . ." McIlhenny (1939), p. 25.

198 "When our tribe was at its lowest ebb . . ." Ibid., p. 57.

200 Details of the selling price of pelts and breeding pairs of foxes from Prince Edward Island come from "U.S. Fox Farming a $50,000 Industry" (1936), p. 176.

201n Information about foxes released on Alaskan islands comes from Bailey (1993), p. 11.

201 "A Ten-Inch Rodent . . ." "The Rare Chinchilla Bred in Captivity" (1934), p. 203.

203 "the mother is doing nicely" Dozier to Ashbrook, January 29, 1940. Smithsonian Archives.

203 Details of the hurricane and the days leading up to it were taken from the *New Orleans Times-Picayune* for August 3–9, 1940.

207 The statistic that Louisiana trappers took 8,786 nutrias in 1945 and 18,015 in 1946 comes from Ashbrook (1948), p. 91.

207 The statistic that in 1961 one million nutrias lived in Louisiana comes from Evans (1970), p. 11.

208 The details about the ways laws changed in Louisiana regarding nutrias comes from ibid., pp. 49–50.

208 The statistic that in 1965–66 trappers made three million dollars off nutrias comes from ibid., p. 58.

209 "It is pertinent at this point . . ." National Better Business Bureau (1957), p. 3.

210 "The epidemic of booms and busts . . ." Ibid., pp. 7–8.

211 The prices for nutria fur versus chinchilla fur come from "Price Schedule" (1999).

CHAPTER FIFTEEN

214 "great mass of sticks" Darwin (1989), p. 139.

216 For an extensive look at parrots as pets in sixteenth-century Europe and as indicators of the exotic, see Boehrer (1998).

217 The statistics on numbers of birds generally and monk parakeets specifically imported to the United States in 1968 come from Banks (1970), pp. 1–26.

217 The statistics on numbers of birds generally and monk parakeets specifically imported to the United States in 1971 come from Clapp and Banks (1973), pp. 14–19.

217 The statistics on numbers of parrots exported from neotropical countries and imported to the United States between 1982 and 1988 come from Bessinger and Snyder (1992), p. 221.

218 The records of numbers and kinds of birds on the White House lawn counted by President Roosevelt in 1908 and the executive director of the Wildlife Society in 1969 come from Robertson (1969), p. 1.

219 The details of the "retrieval" program for monk parakeets come from Neidermyer and Hickey (1977), p. 275.

222 The statistics on the increase of monk parakeets in Hyde Park between 1992 and 1993 come from Hyman and Pruett-Jones (1995), p. 512.

222 The statistics on the number of monk parakeets in Hyde Park in 1997 come from Pruett-Jones and Tarvin (1998), pp. 55–58.

227 "very often, in Autumn . . ." William Byrd, quoted in Wright (1912), p. 347.

227 "The paroquet is found . . ." H. R. Schoolcraft, quoted in ibid., p. 356.

CHAPTER SIXTEEN

238 "second-best friend is a chemophobic environmentalist." Williams (1997), p. 26.

238 "Given the poor knowledge . . ." Simberloff (1996), p. 1972.

240 "brim with meadow crane's bill . . ." Candida Lycett-Green, quoted in "Traveling to Find Yourself" (1997), p. 4.

CHAPTER SEVENTEEN

245 "There may come a time . . ." Auld (1892), p. 169.

BIBLIOGRAPHY

Abs, Michael, ed. 1983. *Physiology and Behavior of the Pigeon.* London: Academic Press.

Adler, Tina. 1994. "Squelching Gypsy Moths." *Science News,* vol. 145, March 19, 1994.

"Advisor to the King of Korea." *Sunday Oregonian Magazine,* August 27, 1950.

Aitken, Hugh G. H. 1954. *The Welland Canal Company: A Study in Canadian Enterprise.* Cambridge: Harvard University Press.

Allen, J. A. 1899. "Economic Relations of Birds to Agriculture." *Auk,* vol. 16, no. 3.

———. 1900. "Economic Ornithology." *Auk,* vol. 17, no. 4.

American Acclimatization Society. 1871. *Charter and By-Laws of the American Acclimatization Society,* New York: Geo. W. Averell.

"American Silk Manufacture." *Scientific American,* vol. 19, no. 8, October 28, 1868.

Andreadis, Theodore G., and Ronald M. Weseloh. 1990. "Discovery of *Entomophaga maimaiga* in North American Gypsy Moth, *Lymantria dispar.*" *Proceedings of the National Academy of Sciences of the United States,* vol. 87, no. 7, April 1990.

Andrews, W. V. 1871. "Silk Culture." *Scientific American,* vol. 24, no. 12, March 18, 1871.

Ashbrook, Frank G. 1922. "The Fur Trade and Fur Supply." *Journal of Mammology,* vol. 3, no. 1.

———. 1948. "Nutrias Grow in the United States." *Journal of Wildlife Management,* vol. 12, no. 1.

"Astronomical Plates." *New York Times,* August 22, 1872.

Atwood, Earl L. 1950. "Life History Studies of Nutria, or Coypu, in Coastal Louisiana." *Journal of Wildlife Management,* vol. 14, no. 3.

Auld, Robert C. 1892. "The Economic Introduction of the Kangaroo in America." *Overland Monthly*, series II, vol. 20.

Averill, A. B. 1895. "The Denny Pheasant." *Oregon Naturalist*, vol. 2, no. 5, May 1895.

Bailey, Edgar P. 1993. *Introduction of Foxes to Alaskan Islands—History, Effects on Avifauna, and Eradication*. U.S. Department of the Interior, Fish and Wildlife Service, Resource Publication no. 193.

Bailey, Solon I. 1931. *History and the Work of the Harvard Observatory, 1839 to 1927*. New York: McGraw Hill Book Co.

Banks, Richard C. 1970. "Birds Imported into the United States in 1968." U.S. Department of the Interior, Fish and Wildlife Service, Bureau of Sport Fisheries and Wildlife, Special Scientific Report—Wildlife no. 136, September 1970.

Barnes, Jeffrey K. 1988. "Asa Fitch and the Emergence of American Entomology." *New York State Museum Bulletin*, no. 461, 1988.

Barrus, Clara. 1925. *The Life and Letters of John Burroughs*. Boston: Houghton Mifflin Co.

Barton, Benjamin Smith. 1783. *An Inquiry into the Question, Whether the Apis Mellifica, or True Honeybee, Is a Native of America*. Philadelphia.

Batra, Suzanne W. T. 1984. "Solitary Bees." *Scientific American*, vol. 250, February 1984.

Baurmeister, Carl Leopold. 1973. *Revolution in America, Confidential Letters and Journals 1776–1784 of Adjutant General Major Baurmeister of the Hessian Forces*. Trans. and annotated by Bernhard A. Uhlendorf. Westport: Greenwood Press.

Beard, Jonathan D. 1991. "Bug Detectives Crack the Tough Cases." *Science*, vol. 245. December 13, 1991.

Beebe, William. 1990. *A Monograph of the Pheasants*. New York: Dover Publications.

Bender, Norman J. 1996. *Winning the West for Christ: Sheldon Jackson and Presbyterianism on the Rocky Mountain Frontier, 1869–1880*. Albuquerque: University of New Mexico Press.

Berger, Andrew J. 1972. *Hawaiian Birdlife*. Honolulu: University Press of Hawaii.

Bessinger, Steven R., and Noel F. R. Snyder. 1992. *New World Parrots in Crisis*. Washington, D.C.: Smithsonian Institution Press.

Bird and Mammal Laboratories, United States Fish and Wildlife Service, circa 1885–1971. Records. Smithsonian Archives.

Bishop, Morris. 1949. *Champlain, The Life of Fortitude*. London: Macdonald.

Blodgett, Howard. 1973. "New Wild Birds in Michigan." *American Cage Bird Magazine*, vol. 45, no. 6, June 1973.

Boehrer, Bruce. 1998. " 'Men, Monkeys, Lap-Dogs, Parrots, Perish All!' Psittacine Articulacy in Early Modern Writing." *Modern Language Quarterly*, vol. 59, no. 2, June 1998.

Bourdelle, E. 1939. "American Mammals Introduced into France in the Contemporary Period, Especially Myocastor and Ondatra." *Journal of Mammology*, vol. 20, no. 3.

Braislin, W. C. 1898. "The Starling (*Sturnus vulgaris*) on Long Island." *Auk*, vol. 15, no. 1.

"Breeding Silkworms." *Scientific American*, vol. 25, no. 17, October 21, 1871.

Brewster, William. 1890. "The Heath Hen." *Forest and Stream*, vol. 35, no. 10, September 25, 1890.

Brotman, Barbara. 1988. "Parrot Troopers Defend Their Feathered Friends." *Chicago Tribune*, April 19, 1988, section 2.

Bryant, William Cullen. 1910. "The Old-World Sparrow." *The Poetical Works of William Cullen Bryant.* New York: D. Appleton and Co.

Buchmann, Stephen L., and Nabhan, Gary Paul. 1996. *The Forgotten Pollinators.* Washington, D.C.: Island Press.

Bull, John. 1973. "Exotic Birds in the New York City Area." *Wilson Bulletin*, vol. 85, no. 4, December 1973.

———, and Edward R. Ricciuti. 1974. "Polly Want an Apple?" *Audubon*, vol. 76, no. 3, May 1974.

Butz Huryn, Vivian M. 1997. "Ecological Impacts of Introduced Honey Bees." *Quarterly Review of Biology*, vol. 72, no. 3, September 1997.

"Buys 85,000 Acres for Bird Preserve." *New York Times*, October 4, 1914.

Caltagirone, L. E., and R. L. Doutt. 1989. "The History of the Vedalia Beetle Importation to California and Its Impact on the Development of Biological Control." *Annual Review of Entomology*, vol. 34.

Champlain, Samuel de. 1907. *Voyages of Samuel de Champlain, 1604–1618.* Ed. W. L. Grant. New York: Charles Scribner's Sons.

Chang, Kenneth. 1997. "Nature's Immigrants." *Newsday*, June 17, 1997.

Chapman, F. M. 1906. "List of Birds Found within Fifty Miles of the American Museum of Natural History." *American Museum Journal*, vol. 6, no. 3.

———. 1925. "The European Starling as an American Citizen." *Natural History*, vol. 25, no. 5.

———. 1940. *Birds of Eastern North America.* New York: D. Appleton-Century Co.

"Charles Valentine Riley." *Scientific American*, September 28, 1895.

Churchill, Frank. 1906. *Reports on the Condition of Educational and School Service and the Management of Reindeer Service in the District of Alaska*. Washington, D.C.: Government Printing Office.

Clapp, Roger B., and Richard C. Banks. 1973. "Birds Imported into the United States in 1971," U.S. Department of the Interior, Fish and Wildlife Service, Special Scientific Report—Wildlife no. 170, October 1973.

Clerke, Agnes M. 1902. *A Popular History of Astronomy during the Nineteenth Century*. London: Adam and Charles Black.

Cohen, Andrew, and James Carlton. 1998. "Accelerating Invasion Rate in a Highly Invaded Estuary." *Science*, vol. 279, January 23, 1998.

Colby, Charles W. 1922. *The Founder of New France*. Toronto: Glasgow, Brook, and Co.

Colden, Cadwallader D. 1825. *Memoir Prepared at the Request of a Committee of the Common Council of the City of New York and Presented to the Mayor of the City at the Celebration of the Completion of the New York Canals*. New York: W. A. Davis.

Conaway, James. 1984. "On Avery Island, Tabasco Sauce Is the Spice of Life." *Smithsonian*, vol. 15.

Conway, Moncure Daniel. 1862. *The Golden Hour*. Boston: Ticknor and Fields.

Coquillett, D. W. 1889."The Imported Australian Ladybird." *Insect Life*, vol. 2, no. 3, September 1889.

Craven, Wesley Frank. 1964. *Dissolution of the Virginia Company*. Gloucester: Peter Smith.

Cronon, William. 1983. *Changes in the Land*. New York: Hill and Wang.

Crosby, Alfred W. 1986. *Ecological Imperialism*. Cambridge: Cambridge University Press.

Dalrymple, Byron W. 1978. *North American Game Animals*. New York: Crown.

Dampier, Robert. 1971. *To the Sandwich Islands on the H.M.S. Blonde*. Honolulu: University Press of Hawaii.

Darwin, Charles. 1967. *On the Origin of Species*. New York: Atheneum.

———. 1989. *Voyage of the Beagle*. London: Penguin Books.

DeBach, P. 1951. "Effects of Insecticides on Biological Control of Insect Pests of Citrus." *Journal of Economic Entomology*, vol. 44.

———, ed. 1973. *Biological Control of Insect Pests and Weeds*. London: Chapman & Hall Ltd.

DeBach, Paul, and David Rosen. 1991. *Biological Control by Natural Enemies.* London: Cambridge University Press.

DeLuca, Thomas H., and Peter Lesica. 1996. "Long-term Harmful Effects of Crested Wheatgrass on Great Plains Grassland Ecosystems." *Journal of Soil and Water Conservation.* vol. 51, no. 5, September–October 1996.

Deutsch, Hermann B. 1939. "The Bird That's Not on Nellie's Hat." *Saturday Evening Post.* vol. 212, October 14, 1939.

Devine, Robert. 1994. "Botanical Barbarians." *Sierra,* vol. 79, no. 1, January–February 1994.

Dewald, Jonathan. 1980. *The Formation of a Provincial Nobility, The Magistrates of the Parliament of Rouen, 1499–1610.* Princeton: Princeton University Press.

Dickinson, Emily. 1960. "254." *The Complete Poems of Emily Dickinson.* Boston: Little, Brown and Co.

"Dirty Dozen." 1996. *U.S. News & World Report,* vol. 121, no. 17, October 28, 1996.

Doutt, R. L. 1967. "Vice, Virtue, and the Vedalia." *Bulletin of the Entomological Society of America,* vol. 4.

————. 1973. "The Historical Development of Biological Control," in *Biological Control of Insect Pests and Weeds.* Ed. Paul DeBach. London: Chapman & Hall Ltd.

Dow, Charles Mason. 1921. *Anthology and Bibliography of Niagara Falls.* Albany: J. B. Lyon Co.

Dymond, J. R. 1922. "A Provisional List of the Fishes of Lake Erie." *University of Toronto Studies,* Biological Series, vol. 20.

East, Ben. 1949. "Is the Lake Trout Doomed?" *Natural History,* vol. 58, no. 9, November 1949.

Eberhard, Jessica R. 1998. "Breeding Biology of the Monk Parakeet." *Wilson Bulletin,* vol. 110, no. 4, December 1998.

Eccles, W. J. 1983. *The Canadian Frontier, 1534–1760.* Albuquerque: University of New Mexico Press.

Ehrlich, Paul; David Dobkin; and Darryl Wheye. 1988. *The Birder's Handbook.* New York: Simon & Schuster.

Elton, Charles S. 1958. *The Ecology of Invasions by Animals and Plants.* London: Methuen & Co. Ltd.

Essig, E. O. 1931. *A History of Entomology.* New York: Macmillan Co.

Evans, James. 1970. *About Nutria and Their Control.* U.S. Department of the Interior, Bureau of Sport Fisheries and Wildlife, Resource Publication no. 86.

Fausch, Kurt D. 1988. "Tests of Competition between Native and Introduced Salmonids in Streams: What Have We Learned?" *Canadian Journal of Fisheries and Aquatic Sciences*, vol. 45, no. 12, December 1988.

———, and Ray J. White. 1981. "Competition between Brook Trout (*Salvelinus fontinalis*) and Brown Trout (*Salmo trutta*) for Positions in a Michigan Stream." *Canadian Journal of Fisheries and Aquatic Sciences*, vol. 38, no. 10, October 1981.

Fesperman, Dan. 1992. "In Andrew's Wake, a New Wild Kingdom." *Baltimore Sun*, September 20, 1992.

Flammarion, Camille. 1880. *Popular Astronomy*. New York: D. Appleton and Co.

Forbush, Edward H., and Charles H. Fernald. 1896. *The Gypsy Moth*. Boston: Wright and Potter Printing Co.

Force, Peter, ed. 1963. *Tracts and Other Papers*, vols. II and III. Gloucester: Peter Smith.

Forshaw, Joseph M. 1978. *Parrots of the World*. Melbourne: Landsdowne Press.

Foster, J. E.; P. L. Taylor; and J. E. Araya. 1986. *The Hessian Fly*. Purdue University Agricultural Experiment Station Bulletin, no. 502.

"Four Mountain Goats Released on Rocky Cliff above Lake Crescent." *Port Angeles Evening News*. January 2, 1925.

Gent, E. W. 1963. "Virginia: More Especially the South Part Thereof, Richly and Truly Valued," in *Tracts and Other Papers*, vol. III. Ed. Peter Force. Gloucester: Peter Smith.

German Society of Fish Breeders. 1880. Letter dated December 14, 1878, to Spencer Baird. *Report of the Commissioner of the United States Commission of Fish and Fisheries for 1878*. Washington, D.C.: Government Printing Office.

Gill, Frank B. 1995. *Ornithology*. New York: W. H. Freeman and Co.

Goff, M. Lee, and Charles van Riper III. 1980. "Distribution of Mosquitoes (Diptera: Culcidae) on the East Flank of Mauna Loa Volcano." *Pacific Insects*, vol. 22, no. 1, August 29, 1980.

Goodman, Billy. 1991. "Keeping Anglers Happy Has a Price." *BioScience*, vol. 41, no. 5, May 1991.

Gordon, Robert D., and Natalia Vandenburg. 1991. "Field Guide to Recently Introduced Species of Coccinellidae (Coleoptera) in North America, with a Revised Key to North American Genera of Coccinellini." *Proceedings of the Entomological Society of Washington*, vol. 93, no. 4.

Greene, S. H. 1893. "Oregon Pheasants and Quail." *Forest and Stream*, vol. 40, no. 18, May 4, 1893.

Gross, Alfred O. 1928. "The Heath Hen." *Memoirs of the Boston Society of Natural History*, vol. 6, no. 4, May 1928.

Gruber, Ira D. 1972. *The Howe Brothers and the American Revolution.* New York: Atheneum.

Haack. 1882. "A German View of the American Section in the Berlin Fishery Exhibition." *Bulletin of the United States Fish Commission for 1881,* vol. I. Washington, D.C.: Government Printing Office.

Hadfield, Charles. 1968. *The Canal Age.* Newton Abbot: David & Charles Ltd.

Hagen, H. A. 1883. "The Hessian Fly Not Imported from Europe." *Third Report of the United States Entomological Commission.* Washington, D.C.: Government Printing Office.

Hale, William H. 1906. "Summer Meeting of the American Association for the Advancement of Science." *Scientific American*, vol. 95, no. 2, July 14, 1906.

Hallock, Charles. 1876. *Camp Life in Florida: A Handbook for Sportsmen and Settlers.* New York: Forest and Stream Publishing Co.

Hardisty, M. W., and I. C. Potter. 1971. "The General Biology of Adult Lampreys." *The Biology of Lampreys*, vol. 1. Ed. M. W. Hardisty and I. C. Potter. New York: Academic Press.

Hardy, John William. 1973. "Feral Exotic Birds in Southern California." *Wilson Bulletin*, vol. 85, no. 4, December 1973.

Harrison, Gordon. 1978. *Mosquitoes, Malaria, and Man.* New York: E. P. Dutton.

Harting, James Edmund. 1965. *The Ornithology of Shakespeare.* Chicago: Argonaut, Inc.

Hilkevitch, Jon. "Babbling Birds Brave Winnetka's Wilds." *Chicago Tribune*, January 9, 1993.

Hill, David, and Peter Robertson. 1988. *The Pheasant, Ecology, Management, and Conservation.* Oxford: BSP Professional Books.

Hodgson, Robert G. 1938. *Nutria Raising.* Toronto: Fur Trade Journal of Canada.

———. 1949. *Farming Nutria for Profit.* Toronto: Fur Trade Journal of Canada.

Holmgren, Virginia C. 1964. "Chinese Pheasants, Oregon Pioneers." *Oregon Historical Quarterly*, vol. 65, no. 3, September 1964.

Houston, Douglas B., and Edward G. Schreiner. 1995. "Alien Species in National Parks: Drawing Lines in Space and Time." *Conservation Biology*, vol. 9, no. 1.

Houston, Douglas B.; Edward G. Schreiner; and Bruce B. Moorhead. 1994. *Mountain Goats in Olympic National Park: Biology and Management of an Exotic Species.* Washington, D.C.: Department of the Interior, National Park Service.

Howard, L. O. 1925. "Albert Koebele, an Obituary." *Journal of Economic Entomology*, vol. 18.

———. 1930. "A History of Applied Entomology." *Smithsonian Miscellaneous Collections*, vol. 84, pub. 3065.

Howell, A. B. 1943. "Starlings and Woodpeckers." *Auk*, vol. 60, no. 1.

Hubbell, Sue. 1996. "How Taxonomy Helps Us Make Sense of the Natural World." *Smithsonian*, vol. 27, no. 2, May 1996.

Hubbs, C. L., and I. C. Potter. 1971. "Distribution, Phylogeny, and Taxonomy." *The Biology of Lampreys*, vol. 1. Ed. M. W. Hardisty and I. C. Potter. New York: Academic Press.

Hyman, Jeremy, and Stephen Pruett-Jones. 1995. "Natural History of the Monk Parakeet in Hyde Park, Chicago." *Wilson Bulletin*, vol. 107, no. 3.

"Increased Use of Homing Pigeons." 1897. *Scientific American*, vol. 76, no. 1, January 9, 1897.

Ingold, D. J. 1994. "Influence of Nest-Site Competition between European Starlings and Woodpeckers." *Wilson Bulletin*, vol. 106, no. 2.

"Is Competition Red in Tooth and Claw?" *Economist*, vol. 343, no. 8022, June 21, 1997.

Jackson, Donald D. 1984. "Gypsy Invaders Seize New Ground in Their War against Our Trees." *Smithsonian*, vol. 15, May 1984.

———. 1985. "Pursued in the Wild for the Pet Trade, Parrots Are Perched on a Risky Limb." *Smithsonian*, vol. 16, April 1985.

Jackson, John. 1997. *The Welland Canals and Their Communities*. Toronto: University of Toronto Press.

Jackson, Sheldon. 1891. Senate Miscellaneous Documents, serial no. 2821. 2d sess. 51st Congress, vol. 3, doc. no. 39. Washington, D.C.: Government Printing Office: 1891.

———. 1893. *Report on the Introduction of Domestic Reindeer into Alaska*. Washington, D.C.: Government Printing Office.

———. 1894. *Report on the Introduction of Domestic Reindeer into Alaska*. Washington, D.C.: Government Printing Office.

———. 1898. *Eighth Annual Report on Introduction of Domestic Reindeer into Alaska*. Washington, D.C.: Government Printing Office.

———. 1902. *Eleventh Annual Report on Introduction of Domestic Reindeer into Alaska*. Washington, D.C.: Government Printing Office.

Jefferson, Thomas. 1975. "Notes on the State of Virginia," in *The Portable Thomas Jefferson.* New York: Viking Press.

———. 1984. "Writings." New York: Literary Classics of the United States.

Jellison, William L. 1946. "Spotted Skunk and Feral Nutria in Montana." *Journal of Mammology,* vol. 26, no. 4.

Jennings, Diane. 1995. "Fair or Unfair Game?" *Dallas Morning News,* September 2, 1995.

Johnson, Richard F., and Marian Janiga. 1995. *Feral Pigeons.* Oxford: Oxford University Press.

Jones, Bessie Z., and Lyle G. Boyd. 1971. *The Harvard College Observatory: The First Four Directorships, 1839–1919.* Cambridge: Belknap Press.

Judd, Gerrit P., IV. 1960. *Dr. Judd, Hawaii's Friend.* Honolulu: University of Hawaii Press.

Kaplan, J. Kim. 1998. "Conserving the World's Plants." *Agricultural Research,* vol. 46, no. 9.

Kessel, B. 1957. "A Study of the Breeding Biology of the European Starling (*Sturnus vulgaris*) in North America." *American Midland Naturalist,* vol. 58, no. 2.

Kingsbury, Susan, ed. 1933. *Records of the Virginia Company of London,* vol. III. Washington, D.C.: Government Printing Office.

Koebele, Albert. 1890. "Report of the Fluted Scale of the Orange and Its Natural Enemies in Australia." U.S. Department of Agriculture, Division of Entomology, Bulletin 21.

Koebele Collection. California Academy of Sciences Archives.

Lanier, Sidney. 1876. *Florida: Its Scenery, Climate, and History.* Philadelphia: J. B. Lippincott & Co.

Lanman, Charles. 1856. *Adventures in the Wilds of the United States and British American Provinces.* Philadelphia: J. W. Moore.

Larrison, Earl J. 1943. "Feral Coypus in the Pacific Northwest." *Murrelet,* vol. 24, no. 1.

Laycock, George. 1966. *The Alien Animals.* Garden City: Natural History Press.

Legget, Robert F. 1979. *The Seaway.* Toronto: Clark, Irwin & Co.

Leopold, Starker A., and Fraser F. Darling. 1953. *Wildlife in Alaska.* New York: Ronald Press Co.

Lescarbot, Marc. 1968. *History of New France,* vol. III. New York: Greenwood Press.

Lever, Christopher. 1987. *Naturalized Birds of the World.* Essex: Longman Scientific and Technical.

Levi, Wendell M. 1974. *The Pigeon.* Sumter: Levi Publishing Co.

"Life in Brittany." *Appleton's Journal,* vol. 8, no. 2, February 1880.

Long, John L. 1981. *Introduced Birds of the World.* New York: Universe Books.

Lowery, George H. 1951. "Obituaries: Edward Avery McIlhenny." *Auk,* vol. 68, no. 1.

Lyman, Chester S. 1924. *Around the Horn to the Sandwich Islands and California, 1845–1850.* New Haven: Yale University Press.

Mahan, John and Ann Mahan. 1998. *Lake Superior, Story and Spirit.* Gaylord: Sweetwater Visions.

Manchester, H. H. 1916. *The Story of Silk and Cheney Silks.* South Manchester: Cheney Brothers, Silk Manufacturers.

Manning, Richard. 1995. *Grassland.* New York: Viking.

Maples, W. R.; A. B. Brown; and P. M. Hutchins. 1976. "Introduced Monkey Populations at Silver Springs, FL." *Florida Anthropologist,* vol. 29.

Masteller, Mark A., and James A. Bailey. 1988. "Agonist Behavior among Mountain Goats Foraging in Winter." *Canadian Journal of Zoology,* vol. 66, no. 11.

Martin, Jim. "Orient's Rise Predicted by Oregonian on Korean King's Staff." *Oregonian,* June 22, 1986.

Martin, Joseph Plumb. 1962. *Private Yankee Doodle.* New York: Popular Library.

Mather, Fred. 1873a. "Fish Culture." *Forest and Stream,* vol. 1, no. 1, August 14, 1873.

———. 1873b. "Bass in Trout Waters." *Forest and Stream,* vol. 1, no. 6, September 18, 1873.

———. 1874. "Practical Fish Culture." *Forest and Stream,* vol. 2, no. 9, April 9, 1874.

———. 1876a. "Apparatus for Hatching Shad Ova while En Route to New Waters." *Report of the Commissioner of Fish and Fisheries for 1873–4 and 1874–5.* Washington, D.C.: Government Printing Office.

———. 1876b. "Voyage to Bremerhaven, Germany, with Shad." *Report of the Commissioner of Fish and Fisheries for 1873–4 and 1874–5.* Washington, D.C.: Government Printing Office.

———. 1881. "The Old World Nuisance." *Forest and Stream,* vol. 17, no. 3, August 18, 1881.

———. 1886a. "Eggs Received from Foreign Countries at Cold Spring Harbor, New York, and Retained or Forwarded during the Seasons of 1883–4 and 1884–5." *Report of the Commissioner of the United States Commission of Fish and Fisheries for 1884.* Washington, D.C.: Government Printing Office.

———. 1886b. "Work at Cold Spring Harbor, Long Island, during 1883 and 1884."

Report of the Commissioner of Fish and Fisheries for 1884. Washington, D.C.: Government Printing Office.

Matthiessen, Peter. 1964. *Wildlife in America.* New York: Viking Press.

McGuire, Hollister D. 1899. "Pheasant Rearing." *Forest and Stream,* vol. 52, no. 8, February 25, 1899.

McIlhenny, E.A. 1933. *Befo' de War Spirituals.* Boston: Christopher Publishing House.

————. 1935. *The Alligator's Life History.* Boston: Christopher Publishing House.

————. 1936. "Are Starlings a Menace to the Food Supply of Our Native Birds?" *Auk,* vol. 53, no. 3.

————. 1939. *The Autobiography of an Egret.* New York: Hastings House.

Mitchell, G. "Enhancement of Reindeer Herd Health." Alaska Agricultural and Forestry Experiment Station, 1996 [cited December 17, 1998]. Available at http://hermes.ecn.purdue.edu:8001/Links/impact96/0007.html

Montaigne, Michel de. 1927. *The Essays of Montaigne,* vol. I. Trans. E. J. Trechmann. London: Oxford University Press.

Morgan, Theodore. 1948. *Hawaii, a Century of Economic Change, 1778–1876.* Cambridge: Harvard University Press.

Morton, Thomas. 1883. *The New English Canaan.* Boston: Prince Society.

Morill, Wendell. 1991. "Montana Small Grain Insects, Hessian Fly." Montana State University, Bozeman, MT, Capsule Information Series, no. 38, October 1991.

Mss. no. 2134, Oregon Historical Society Archives.

Muir, John. 1988. *Travels in Alaska.* San Francisco: Sierra Club Books.

————. 1993. *Letters from Alaska.* Ed. Robert Engberg and Bruce Merrill. Madison: University of Wisconsin Press.

Munro, George C. 1960. *Birds of Hawaii.* Rutland: Bridgeway Press.

Munro, Keith. 1950. "Slaughter in the Great Lakes." *Reader's Digest,* vol. 56, no. 337, May 1950.

Nairn, Fraser. 1934. "Modern Garden of Eden." *Country Life,* vol. 67.

National Better Business Bureau. Service Bulletin, no. 1602, May 1957.

Neidermyer, William J., and Joseph J. Hickey. 1977. "The Monk Parakeet in the United States, 1970–75." *American Birds,* vol. 31, no. 3, May 1977.

New Orleans Times-Picayune, August 3–9, 1940.

"Newsclips of Sorcerers, Pests, and Downright Ugly Matters in 1787." *Life,* vol. 10, September 22, 1987.

Nyman, O. L. 1970. "Ecological Interaction of Brown Trout, Salmo trutta L., and Brook Trout, Salvelinus fontinalis (Mitchill), in a Stream." *Canadian Field-Naturalist*, vol. 84, no. 4, October–December 1970.

"The Oatmeal Ark." *Economist*, vol. 343, no. 8017. May 17, 1997.

Ober, Frederick A. 1874. "Lake Okechobee, Part I." *Appleton's Journal*, vol. 12, no. 293, October 31, 1874.

Office of Education, Alaska Division, Bureau of Indian Affairs. General Correspondence. National Archives.

Olinger, David. 1995. "Monkey Business Ends in Florida." *St. Petersburg Times*, February 3, 1995.

———. 1997. "Amok on Monkey Island." *St. Petersburg Times*, February 9, 1997.

Olson, Dean F. 1969. *Alaska Reindeer Herdsmen*. Institute of Social, Economic, and Government Research, University of Alaska, Report no. 22.

Olson, S. L., and H. F. James. 1982. "Fossil Birds from the Hawaiian Islands, Evidence for Wholesale Extinction before Western Contact." *Science*, vol. 217, August 13, 1982.

Olympic National Park. 1987. *Mountain Goats in Olympic National Park*. Port Angeles: Olympic National Park.

Onstad, David, and Michael McManus. 1996. "Risks of Host Range Expansion by Parasites of Insects." *BioScience*, vol. 42, no. 6, June 1996.

Owre, Oscar T. 1973. "A Consideration of the Exotic Avifauna of Southeastern Florida." *Wilson Bulletin*, vol. 85, no. 4, December 1973.

Oxford English Dictionary. 1961. London: Oxford University Press.

Page, Jake. 1990. "Pushy and Brassy, the Starling Was an Ill-Advised Import." *Smithsonian*, vol. 21, no. 6, September 1990.

"Paper Shells." *Oregonian*, March 19, 1882.

Perrault, Anna H. ed. 1987. *Nature Classics: A Catalogue of the E. A. McIlhenny Natural History Collections at Louisiana State University*. Baton Rouge: Friends of the LSU Library.

Pettengill, Ray W., trans. 1964. *Letters from America, 1776–1779, Being Letters of Brunswick, Hessian, and Waldeck Officers with the British Armies during the Revolution*. New York: Kennikat Press.

Pierce, Robert A. 1956. "Some Thoughts Concerning the Introduction of Exotic Game Birds." *Wilson Bulletin*, vol. 68, no. 1.

Pliny the Elder. 1991. *Natural History*. Trans. John Healy. London: Penguin Books.

Presnall, Clifford C. 1958. "The Present Status of Exotic Mammals in the United States." *Journal of Wildlife Management*, vol. 22, no. 1.

"Price Schedule." Fur Age [cited May 13, 1999]. Available at http://www.furs.com/price.html

Proctor, Michael; Peter Yeo; and Andrew Lack. 1996. *The Natural History of Pollination*. Portland: Timber Press.

The Prophet Nathan, or, Plain Friend. 1788. Hudson: Ashbel Stoddard.

Pruett-Jones, Stephen, and Keith A. Tarvin. 1998. "Monk Parakeets in the United States: Population Growth and Regional Patterns of Distribution." *Proceedings of the 18th Vertebrate Pest Conference* Eds. R. O. Baker and A. C. Crabb. University of California, Davis.

Putnam, Nina Wilcox. 1941. "The Jungle on the Highway." *Scribner's Commentator*, vol. 9, no. 4, February 1941.

"The Rare Chinchilla Bred in Captivity." *Illustrated London News*, vol. 94, no. 2442, 1934.

Rasmussen, Wayne, ed. 1960. *Readings in the History of American Agriculture*. Urbana: University of Illinois Press.

Ray, Dorothy Jean. 1975. *The Eskimos of the Bering Strait, 1650–1898*. Seattle: University of Washington Press.

————. 1983. "The Bering Strait Eskimo" in *Ethnohistory in the Arctic*. Ed. R. A. Pierce. Kingston: Limestone Press.

Recer, Paul. 1997. "Exotic Weevil Threatens Thistles." Associated Press News Service. August 21, 1997.

"Reciprocal Treaties of Commerce." 1853. *Debow's Review, Agricultural, Commercial, Industrial Progress and Resources*, vol. 14, no. 6, June 1853.

"Reindeer for Alaska." *New York Times*, November 22, 1892.

Renaud, M. G. 1897. "Evolution of the Carrier Pigeon." *Appleton's Popular Science Monthly*, vol. 50, no. 23, January 1897.

Riley, Charles Valentine. 1879. *Report of the Entomologist for the Year 1878*. Washington, D.C.: Government Printing Office.

————. 1889. *Report of the Entomologist for the Year 1888*. Washington, D.C.: Government Printing Office.

————. 1890. *Report of the Entomologist for the Year 1890*. Washington, D.C.: Government Printing Office.

————. 1892. *Report of the Secretary of Agriculture for 1891, Division of Entomology*. Washington, D.C.: Government Printing Office.

Rinhard, Floyd, and Marion Rinhard. 1986. *Victorian Florida.* Atlanta: Peachtree Publishers Limited.

"Robert C. M'c Auld, a Scottish Writer." *New York Times,* April 23, 1937.

Robertson, Archibald. 1971. *Archibald Robertson, His Diaries and Sketches in America, 1762–1780.* Ed. Harry Miller Lydenberg. New York: New York Public Library and Arno Press.

Robertson, Nan. 1969. "Fewer Birds Visiting White House." *New York Times,* December 28, 1969.

Rothschild, Miriam, and Theresa Clay. 1952. *Fleas, Flukes, and Cuckoos: A Study of Bird Parasites.* New York: Philosophical Library.

Russell, Hamlin. 1893. "The Story of the Buffalo." *Harper's New Monthly Magazine,* vol. 86, no. 515.

St. Catherine's Historical Museum. *A Canadian Enterprise: The Welland Canals, The "Merritt Day" Lectures—1978–82.* St. Catherine's: St. Catherine's Historical Museum.

Sammataro, Diana, and Alphonse Avitabile. 1998. *The Beekeeper's Handbook.* Ithaca and London: Cornell University Press.

Samuels, Edward A. 1870. *The Birds of New England and Adjacent States.* Boston: Noyes, Holmes and Co.

Sanders, D. A. 1940. "Rhesus Monkeys (Macaca mulatta) for American Laboratories." *Science,* vol. 92, no. 2387, September 27, 1940.

"Salt-Water Hatchery at Cold Spring Harbor." *Forest and Stream,* vol. 21, no. 12, October 18, 1883.

Saunders, R. M. 1935. "The First Introduction of European Plants and Animals into Canada." *Canadian Historical Review,* vol. 16, no. 4.

Scheffer, V. B. 1951. "The Rise & Fall of a Reindeer Herd," *Scientific Monthly,* vol. 73, no. 6.

———. 1993. "The Olympic Mountain Goat Controversy: A Perspective." *Conservation Biology 4,* vol. 7, no. 4, December 1993.

Schierbeek, A. 1967. *Jan Swammerdam: His Life and Works.* Amsterdam: Swets & Zeitlinger.

Schorger, A.W. 1952. "Introduction of the Domestic Pigeon." *Auk,* vol. 69, no. 4.

Scrapbooks, Oregon Historical Society Archives.

Sears, Hamblen. 1897. "Pheasant and Quail Three Hours from Town." *Harper's Weekly,* vol. 41, no. 2128, October 2, 1897.

Seeley, Thomas D. 1985. *Honeybee Ecology.* Princeton: Princeton University Press.

Shakespeare, William. 1974. *The Riverside Shakespeare.* Boston: Houghton Mifflin Co.

Sharp, Ward M. 1957. "Social and Range Dominance in Gallinaceous Birds—Pheasants and Prairie Grouse." *Journal of Wildlife Management*, vol. 21, no. 2, April 1957.

Shaw, William T. 1908. *The China Pheasant or Denny Pheasant in Oregon.* Philadelphia: J. B. Lippincott Co.

Sheehan, Bernard. 1980. *Savagism and Civility: Indians and Englishmen in Colonial Virginia.* Cambridge: Cambridge University Press.

"Silk and Its Culture." *Scientific American*, vol. 19, no. 3, July 15, 1868.

"Silk Manufactures of Lyons." *Scientific American*, vol. 18, no. 26, June 27, 1868.

"Silk Reeling in California." *Scientific American*, vol. 25, no. 19, November 4, 1871.

Simberloff, Daniel, and Peter Stiling. 1996. "How Risky Is Biological Control?" *Ecology*, vol. 77, no. 7.

Simberloff, Daniel; Don C. Schmitz; and Tom C. Brown. 1997. *Strangers in Paradise.* Washington, D.C.: Island Press.

Simpson, J. A., and E. S. C. Weiner, eds. 1989. *The Oxford English Dictionary*, 2d ed., vol. XVI. Oxford: Oxford University Press.

Singer, Charles. 1959. *A History of Biology to about the Year 1900.* London and New York: Abelard-Schuman.

Singer, Daniel J. 1925. *Big Game Fields of America, North and South.* New York: George H. Doran Company.

Smith, Bernard, R. 1971. "Sea Lampreys in the Great Lakes of North America." *The Biology of Lampreys*, vol. 1. Ed. M. W. Hardisty and I. C. Potter. New York: Academic Press.

Smith, J. B. 1910. "Insects and Entomologists, Their Relation to the Community at Large, II." *Popular Science Monthly*, vol. 76, no. 29, May 1910.

Smith, John. 1907. "Description of Virginia and Proceedings of the Colonie," in *Narratives of Early Virginia.* Ed. Lyon Tyler. New York: Charles Scribner's Sons.

Snow, Keith, R. 1990. *Mosquitoes.* Slough: Richmond Publishing Co.

"Social Life in the Tropics." *Overland Monthly*, vol, 1, no. 6, December 1868.

Staletovich, Jenny. 1997. "Keys-Killing Monkeys Must Go." *Palm Beach Post*, July 10, 1997.

Stewart, Robert Laird. 1908. *Sheldon Jackson: Pathfinder and Prospector of the Mis-*

sionary Vanguard in the Rocky Mountains and Alaska. New York: Fleming H. Revell Co.

Stoll, Steven. 1995. "Insects and Institutions: University Science and the Fruit Business in California." *Agricultural History*, vol. 69, no. 2, Spring 1995.

Stolz, Judith, and Judith Schnell, eds. 1991. *Trout.* Harrisburg: Stackpole Books.

Stone, Livingston. 1876. "Report of Operations in California." *Report of the Commissioner for Fish and Fisheries for 1873–4 and 1874–5.* Washington, D.C.: Government Printing Office.

Story, J. M. 1992. "Biological Control of Weeds: Selective, Economical, and Safe." *Western Wildlands*, Summer 1992.

———. 1996. "Refuge Is Site for Biological Weed Control Study." *The Teller, Notes from Teller Wildlife Refuge*," vol. 3, no. 2, Spring 1996.

———; W. R Good; and L. J. White. 1994. "Propagation of *Agapeta zoegana* L. (Lepidoptera: Cochylidae) for Biological Control of Spotted Knapweed: Procedures and Cost." *Biological Control 4.* New York: Academic Press, Inc.

———; Robert M. Nowierski; and Keith W. Boggs. 1987. "Distribution of *Urophora affinis* and *U. quadrifasciata*, Two Flies Introduced for Biological Control of Spotted Knapweed (*Centaurea maculosa*) in Montana." *Weed Science*, vol. 35.

Swartout, Robert. *Mandarins, Gunboats, and Power Politics.* Honolulu: University Press of Hawaii, 1980.

Syroechkovskii, E. E. 1995. *Wild Reindeer.* Washington, D.C.: Smithsonian Institution Libraries.

Tarr, Cheryl L., and Robert C. Fleischer. 1993. "Mitochondrial-DNA Variation and Evolutionary Relationships in the Amakihi Complex." *Auk*, vol. 110, no. 4.

Taylor, Albert P. 1899. "Pigeongram Service." *Harper's Weekly*, vol. 43, no. 2233, October 7, 1899.

Taylor, Susan Champlin. 1998. "Searching for Hope in the Family Tree." *National Wildlife*, vol. 36, no. 3, April–May 1998.

Tennyson, Alfred Lord. 1899. *Tennyson's Poetical Works.* Boston and New York: Houghton Mifflin Co.

Thorndyke, Lynn. 1958. *A History of Magic and Experimental Science*, vol. VIII. New York: Columbia University Press.

Townsend, Charles W. 1915. "Notes on the Rock Dove (Columba Domestica)." *Auk*, vol. 32, no. 3.

Trimm, Wayne. 1972. "The Monk Parrot." *The Conservationist*, vol. 26, no. 6, June–July 1972.

———. 1973. "Monk Parrots—A Year Later." *The Conservationist,* vol. 27, no. 6, June–July 1973.

Trouvelot, Etienne L. 1867. "The American Silkworm." *American Naturalist,* vol. 1, no. 2, April 1867.

———. 1876a. "On the Veiled Solar Spots." *American Journal of Science and Arts,* vol. XI.

———. 1876b. "On Some Physical Observations of the Planet Saturn." *American Journal of Science and Arts,* vol. XI.

———. 1878a. "The Moon's Zodiacal Light." *American Journal of Science and Arts,* vol. XV.

———. 1878b. "Sudden Extinction of the Light of a Solar Protuberance." *American Journal of Science and Arts,* vol. XV.

———. 1878c. "Undulations Observed in the Tail of Coggia's Comet." *American Journal of Science and Arts,* vol. XV.

———. 1878d. "Observations of the Transit of Mercury, May 5–6." *American Journal of Science and Arts,* vol. XVI.

———. 1882. *The Trouvelot Astronomical Drawings Manual.* New York: Charles Scribner's Sons.

Tyler, Lyon, ed. 1907. *Narratives of Early Virginia.* New York: Charles Scribner's Sons.

U.S. Department of Agriculture. 1883. "The Hessian Fly—Its Ravages, Habits, and the Means of Preventing Its Increase." *Third Report of the United States Entomological Commission.* Washington, D.C.: Government Printing Office.

U.S. Department of the Interior. 1933. *Survey of the Alaska Reindeer Service 1931–1933.* Washington, D.C.: Department of the Interior, Office of the Secretary.

U.S. Department of the Interior, National Park Service. 1987. *The Management of Mountain Goats in Olympic National Park: Some Questions and Some Answers.* Port Angeles: Department of the Interior.

U.S. Fish and Wildlife Service. 1970. *Controlling: Crop Depredations of Blackbirds, Starlings.* Washington, D.C.: Department of the Interior.

"U.S. Fox Farming a $50,000 Industry." *Fortune,* vol. 14, no. 6, 1936.

Van Bael, Sunshine, and Stephen Pruett-Jones. 1996. "Exponential Growth of Monk Parakeets in the United States." *Wilson Bulletin,* vol. 108, no. 3.

Van Riper, Charles, III. 1987. "Breeding Ecology of the Common Amakihi." *Condor,* vol. 89, no. 1.

———, and Sandra G. Van Riper. 1986. "The Epizootiology and Ecological Significance of Malaria in Hawaiian Land Birds." *Ecological Monographs*, vol. 56, no. 4.

Vellequette, Larry P. 1996. "Ohio Town Tries to Get a Drop on Its Pesky Bird Problem." *Toledo Blade*, January 4, 1996.

Virgil. 1957. *The Georgics of Virgil.* Trans. C. Day Lewis. London: Jonathan Cape.

Wagenvoord, Helen Carolyn. 1995. "Mountain Goats in Olympic National Park: Fair Game or Reasonable Doubt?" Thesis (M.S.), University of Montana, Missoula.

Wagner, Balthasar. 1883. "Observations on the New Crop Gall-Gnat." *Third Report of the United States Entomological Commission*. Washington, D.C.: Government Printing Office.

Walcott, Frederic C. 1945. "Historical Introduction," in *The Ring-Necked Pheasant and Its Management in North America*. Washington, D.C.: American Wildlife Institute.

Walton, Perry. 1912. *The Story of Textiles*. Boston: Walton Advertising and Printing Company.

Warner, Richard E. 1968. "The Role of Introduced Diseases in the Extinction of the Endemic Hawaiian Avifauna." *Condor*, vol. 70, no. 2, May 1968.

Washington, George. 1925. *The Diaries of George Washington, 1748–1799*. Boston: Houghton Mifflin Co.

Waters, Thomas F. 1983. "Replacement of Brook Trout by Brown Trout over 15 Years in a Minnesota Stream: Production and Abundance." *Transactions of the American Fisheries Society*, vol. 112, no. 2A, March 1983.

Webster, E. B., ed. 1918. *The Klahhane Annual*. Port Angeles: Klahhane Club.

Webster, E. B. 1920. *The King of the Olympics: The Roosevelt Elk and Other Mammals of the Olympic Mountains*. Port Angeles: Evening News, Inc.

———. 1921. *The Friendly Mountain*. Port Angeles: Evening News, Inc.

———. 1923. *Fishing in the Olympics*. Port Angeles: Evening News, Inc.

———. 1925. "Introduction of Mountain Goats in the Olympic Mountains. *Murrelet*, vol. VI, no. 1, January 1925.

———. 1932. "Status of Mountain Goats Introduced into the Olympic Mountains, Washington." *Murrelet*, vol. XIII, January 1932.

Weitzel, Norman. 1987. "Nest-Site Competition between European Starlings and Native Breeding Birds in Northwestern Nevada." *Condor*, vol. 90, no. 2, May 1987.

Weseloh, Ronald M. 1998. "Possibility for Recent Origin of the Gypsy Moth (Lepidoptera: Lymantriidae) Fungal Pathogen *Entomophaga maimaiga* (Zygomycetes: Entomophthorales) in North America." *Environmental Entomology,* vol. 27, no. 2.

West, M. J. 1990. "Mozart's Starling." *American Scientist,* vol. 78, no. 2.

Westemeier, Ronald L. 1998. "Parasitism of Greater Prairie Chicken Nests by Ring-Necked Pheasants." *Journal of Wildlife Management,* vol. 62, no. 3.

Wilcove, David, et al. 1998. "Quantifying Threats to Imperiled Species in the United States." *BioScience,* vol. 48, no. 8, August 1998.

Williams, Ted. 1997. "Killer Weeds." *Audubon,* vol. 99, no. 2, March–April 1997.

Wilson, Samuel T. 1825. "Ode for the Grand Canal Celebration at the New York Coffee House." *New York Statesman,* November 5, 1825.

Wing, L. 1943. "Spread of the Starling and English Sparrow." *Auk,* vol. 60, no. 1.

Wister, Owen. 1893. "The White Goat and His Country" in *American Big Game Hunting: The Book of the Boone and Crockett Club.* New York: Forest and Stream Publishing Co.

Wohlschke-Bulmahn, Joachim. 1997. "Garden Variety Xenophobia." *Harper's Magazine,* vol. 294, no. 1761, February 1997.

Wolfe, Linda D., and Elizabeth H. Peters. 1987. "History of the Free Ranging Rhesus Monkeys (*Macaca mulatta*) of Silver Springs." *Florida Scientist,* vol. 50, no. 4.

Wolfheim, Jaclyn H. 1983. *Primates of the World, Distribution, Abundance, Conservation.* Seattle: University of Washington Press.

Woolfe, Henry D. 1891. "The Yukon River Explorer Testifies." *In the Senate of the United States, Mr. Teller Presented the Following Newspaper Communication of Sheldon Jackson.* Washington, D.C.: Government Printing Office.

Wright, Albert Hazen. 1912. "Early Records of the Carolina Paroquet." *Auk,* vol. 29, no. 3.

Wright, Karen. 1998. "Building a Better Bee." *New York Times Magazine,* May 10, 1998.

Zorpette, Glenn. 1997. "Parakeets and Plunder." *Scientific American,* vol. 277, no. 1, July 1997.

Zumbo, Jim, and Mike Strandlund. 1989. *Western Big Game Hunting.* National Rifle Association of America.

ACKNOWLEDGMENTS

I AM indebted to the friends who read chapters, offered suggestions, tracked down obscure facts in libraries near where they lived, or spotted an article about mitten crabs, yellow star thistle, or Japanese long-horned beetles and sent it to me in the mail. They constitute a network of support that spans both coasts. Hank Harrington, Don Snow, and Deirdre McNamer at the University of Montana offered invaluable comments and encouragement.

The PEN/Jeràrd Fund, the Erasmus Scholarship and Bertha Morton Scholarship at the University of Montana, a travel grant from the Environmental Studies Department, and the Guerry Beam Fellowship to the High Country Institute for the Study of Journalism and Natural Resources provided funds that allowed me to work on this project.

Librarians at the Mansfield Library, the Smithsonian Archives, the Oregon Historical Society Archives, the National Archives, and the California Academy of Sciences Archives were very helpful. I also greatly appreciate the scientists who were willing to talk to me about their work, particularly Stephen Pruett-Jones at the University of Chicago and Jim Story, Bill Good, and Linda White at Montana State University's Western Agricultural Research Center.

Finally, to Peter and Gail Todd, whose loves for the world of science and the world of words have inspired me in equal measure; to Tamar Todd, who reminds me what's important; and to Jay Stevens, whose keen eye and intelligence make him an excellent reader and whose good heart makes him a great husband, thank you.

INDEX